Umwelt
66,- €
'19

Applied Environmental Science and Engineering for a Sustainable Future

Series editors

Applied Environmental Science and Engineering for a Sustainable Future (AESE) series covers a variety of environmental issues and how they could be solved through innovations in science and engineering. Our societies thrive on the advancements in science and technology which pave the way for better standard of living. The adverse effect of such improvements is the deterioration of the environment. Thus, better catchment management in order to sustainably manage all types of resources (including water, minerals and others) is of paramount importance. Water and wastewater treatment and reuse, solid and hazardous waste management, industrial waste minimisation, soil restoration and agriculture as well as myriad of other topics needs better understanding and application. This book series aims at fulfilling such a task in coming years.

More information about this series at http://www.springer.com/series/13085

Veeriah Jegatheesan • Ashantha Goonetilleke
John van Leeuwen • Jaya Kandasamy
Doug Warner • Baden Myers
Muhammed Bhuiyan • Kevin Spence
Geoffrey Parker
Editors

Urban Stormwater and Flood Management

Enhancing the Liveability of Cities

Springer

Editors
Veeriah Jegatheesan
RMIT University
Melbourne, VIC, Australia

Ashantha Goonetilleke
Queensland University of Technology
Brisbane, QLD, Australia

John van Leeuwen
University of South Australia
Adelaide, SA, Australia

Jaya Kandasamy
University of Technology Sydney
Ultimo, NSW, Australia

Doug Warner
AERU
Hatfield, Hertfordshire, UK

Baden Myers
University of South Australia
Adelaide, SA, Australia

Muhammed Bhuiyan
School of Engineering
Royal Melbourne Institute of Technology
Melbourne, VIC, Australia

Kevin Spence
Department of the Natural and Built Environment
Sheffield Hallam University
Sheffield, UK

Geoffrey Parker
School of Civil Engineering
Liverpool John Moores University
Liverpool, UK

ISSN 2570-2165 ISSN 2570-2173 (electronic)
Applied Environmental Science and Engineering for a Sustainable Future
ISBN 978-3-030-11817-4 ISBN 978-3-030-11818-1 (eBook)
https://doi.org/10.1007/978-3-030-11818-1

Library of Congress Control Number: 2019935170

This Springer imprint is published by the registered company Springer Nature Switzerland AG.
The registered company address is: Gewerbestrasse 11, 6330 Cham, Switzerland

Preface

The liveability of our cities is being compromised more frequently by extreme weather events. The water environment is also being degraded significantly. Furthermore, cities are increasingly water stressed, resulting in costly and reactive strategies such as resorting to seawater and inland brackish water desalination. This is attributed to the growing urbanisation and escalating water demand which are compounded by changing climate conditions such as during strong El Niño and La Niña events that significantly impact on rainfall patterns. The consequences can include significant environmental degradation and increasing economic and social burden on the affected communities. Stormwater management and flood mitigation initiatives such as water-sensitive urban design (WUSD), sponge city and sustainable urban drainage systems (SuDS) are being increasingly implemented to overcome these growing challenges.

However, implementation of such strategies can be localised and may not be integrated with the wider geophysical and social characteristics of the broader surrounding region. There are highly varied approaches towards stormwater harvesting, ranging from advanced systems such as aquifer storage, transfer and recovery to simple rainfall capture from roofs, with technologies and new approaches in ongoing development. This book addresses some of the current and likely challenges in stormwater management that may occur in the future such as increased risks to water quality, environmental impacts and the use of stormwater as a resource for human and environmental needs. Stormwater management challenges are further intensified by emerging and recently identified problems such as transport of key synthetic organic pollutants, e.g. Per- and poly-fluoroalkyl substances (PFAS), antibiotic-resistant bacteria and genes, transport of nutrients and social acceptance and current regulations relating to stormwater capture and reuse. Future impacts of growing urbanisation on flood risks, also associated with changing climate and mitigation, need to be considered as well. The effective management of stormwater in our cities not only addresses many of these challenges but also can enhance the quality of life, local biodiversity and other environmental attributes, as well as the health and well-being of residents. This book explores how responses to all these considerations can be better integrated to enhance the liveability of our cities.

Now, how did this book idea come about? Meetings between Australian Technology Network (ATN) of universities (comprising of 5 universities in Australia) and the University Alliance of the UK (comprising of 19 universities in the UK) identified research areas and knowledge dissemination on stormwater harvesting and flood management and requested interested members to form a consortium to apply for an ATN Science and Research Priorities Seed Fund. Four out of five ATN Network of Universities (Queensland University of Technology, RMIT University, University of South Australia and University of Technology Sydney) met and agreed to conduct a review to identify current practices related to stormwater and flood management as well as to identify the research needs and industry imperatives in this field. The proposal was approved for funding in November 2016. The ATN partners of the project communicated with researchers from the University Alliance of the UK (University of Hertfordshire, Sheffield Hallam University, University of Portsmouth, Nottingham Trent University and Liverpool John Moores University), and the University of Hertfordshire and Sheffield Hallam University formed a consortium with the ATN partners of this project to contribute to a book. Thus, this book is a culmination of several meetings in Australia and a meeting in the UK and from subsequent discussions among the research partners.

This book comprises eight chapters, starting with an introductory chapter that discusses past and present stormwater practices around the globe. It discusses the various initiatives across the world such as WSUD, sponge city programme, SuDS and low impact development (LID). Stormwater governance in Australia, Baltic Sea Region, Canada and the USA, Latin America, the Republic of South Africa and Southeast Asia is also briefly addressed, including case studies.

The second, fourth and sixth chapters of this book deal with UK perspectives on stormwater harvesting and flood mitigation. These include an overview of the issues and barriers to potential solutions (Chap. 2) followed by an examination of WSUD (Chap. 4) and stormwater harvesting (Chap. 6) in detail. Reliance on centralised water supply systems and the lack of small-scale locally implementable techniques are discussed in Chap. 2. Appropriate WSUD systems for different landscapes and their design practice, field implementation and life cycle assessment are discussed in Chap. 4. Factors to be considered in stormwater harvesting such as pollution, treatment, constraints to wider adoption and environmental and economic considerations are included in Chap. 6. Chapter 5 deals with the Australian perspectives on recycling and treatment of stormwater under urban intensification. Several case studies are provided as exemplars of successful implementation of rainwater harvesting, treatment of stormwater through wetlands and managed aquifer storage and recharge. The use of treated wastewater is also included to show the potential of stormwater usage for similar purposes. It is important to know the types of pollutants present in stormwater and their fate before delving into WSUD and stormwater recycling. Accordingly, Chap. 3 provides a detailed overview of urban stormwater quality. Types of pollutants and their concentrations, transport processes as well as their variability and uncertainty, impacts and knowledge gaps are discussed in this chapter.

Both the Australian and UK perspectives on urban stormwater and flood management are further analysed in Chaps. 7 and 8. While Chap. 7 focusses on the challenges, the governance and the knowledge gaps and barriers in implementing urban stormwater and flood management, Chap. 8 considers the relationship between biodiversity and ecosystem services and urban stormwater and flood management.

We hope that this book provides a valuable compilation and consolidation of information that facilitates improved understanding of stormwater management and flood mitigation based on Australian and UK perspectives. A primary aim in producing this book was that it would form the basis for the development of a framework for implementation of integrated and optimised stormwater management strategies in order to mitigate the adverse impacts of an expanding urban water footprint.

Melbourne, VIC, Australia	Veeriah Jegatheesan (Jega)
Brisbane, QLD, Australia	Ashantha Goonetilleke
Adelaide, SA, Australia	John van Leeuwen
Ultimo, NSW, Australia	Jaya Kandasamy
Hatfield, Hertfordshire, UK	Doug Warner
Adelaide, SA, Australia	Baden Myers
Melbourne, VIC, Australia	Muhammed Bhuiyan
Sheffield, UK	Kevin Spence
Liverpool, UK	Geoffrey Parker

Contents

About the Editors and Contributors

Editors

Jega Jegatheesan is Professor in Environmental Engineering and the Director of Water: Effective Technologies and Tools (WETT) Research Centre at RMIT University, Australia. Jega has conducted extensive research on the application of membrane bioreactors, sugar cane juice clarification, seawater desalination and water and stormwater pollution and treatment.

Ashantha Goonetilleke is Professor in Water/Environmental Engineering at Queensland University of Technology (QUT). His areas of expertise include stormwater pollution and mitigation, water-sensitive urban design (WSUD), climate change adaptation, integrated water resources management (IWRM) and stormwater and wastewater recycling. He is currently the Discipline Leader for Civil Engineering in Science and Engineering Faculty, QUT.

John van Leeuwen is Professor in Water Science and Engineering at the University of South Australia. John has ~ 12 years' experience working in the water industry (South Australian Water Corporation) and about 12 years as an academic (University of South Australia). His research is in the areas of drinking water treatment optimisation and modelling, domestic and industrial (pulp and paper mill) wastewater treatment and modelling, water quality processes in catchments and stormwater quality.

Jaya Kandasamy is Associate Professor in Water Engineering at the University of Technology Sydney. He has worked in the industry from 1990 until 2005 mostly in the NSW Government. He was the government's principal reviewer for technical reports funded under its Floodplain Management Program. His areas of expertise include hydraulics, water and stormwater quality and water and stormwater treatment.

Doug Warner is a Senior Researcher and Lecturer within AERU, University of Hertfordshire. Main research interests include ecosystem service provision, carbon footprinting and greenhouse gas mitigation, life cycle analysis, environmental impact and sustainability assessment. Also, he has experience in non-chemical methods of weed control on hard surfaces and habitat management to enhance biodiversity in rural and urban environments.

Baden Myers is a Research Engineer at the University of South Australia. Baden graduated from the University of Technology, Sydney, with a Bachelor of Engineering (Civil and Environmental) in 2005. He completed his PhD based on a study of the physical, chemical and biological processes affecting stormwater quality in permeable pavement structures. His current research interests include stormwater harvesting, urban hydrology, green infrastructure, stormwater drainage and water-sensitive urban design (WSUD).

Muhammed Bhuiyan is a Senior Lecturer in Water Engineering, RMIT University. He has more than 20 years of research and teaching experience in the field of water resources. His areas of expertise include stormwater drainage and flood control systems, climate change, mathematical modelling, river morphology, groundwater management, geostatistics, drought analysis, evapotranspiration simulation, process fluid mechanics and water management in irrigation.

Kevin Spence is a Senior Lecturer at Sheffield Hallam University. Kevin's teaching interests include domestic house construction (historic and modern), building conservation, land surveying, structures, geotechnics, water engineering, flood estimation and statistics. His research interests focus on solute and sediment transport and its impact on water quality in urban drainage and river systems during steady flows and rainfall events using laboratory and field investigative techniques, traditional building construction and land evaluation in a Mediterranean climate.

Geoffrey Parker is a Senior Lecturer in Water Engineering at Liverpool John Moores University. He is currently the Principle of BioFlume which specialises in the development of novel and disruptive solutions for the optimisation/protection of infrastructure and natural resource assets. He is also a contributor and named researcher on several major grants, including the successful 2014 (~ £4.5M) award for the Centre for Doctoral Training (CDT) in Future Infrastructure and the Built Environment (FIBE) funded by the UK EPSRC.

Contributors

Rupak Aryal School of Civil and Environmental Engineering, University of Technology, Sydney, Australia

John Awad Natural and Built Environments Research Centre, School of Natural and Built Environments, University of South Australia, Mawson Lakes, SA, Australia

Public Works Department, Faculty of Engineering, Mansoura University, Mansoura, Egypt

Andrew S. Ball School of Science, RMIT University, Melbourne, VIC, Australia

Jonathan Bridge Department of the Natural and Built Environment, Sheffield Hallam University, Sheffield, UK

Lynn Crowe Department of the Natural and Built Environment, Sheffield Hallam University, Sheffield, UK

Ashantha Goonetilleke School of Civil Engineering and Built Environment, Queensland University of Technology (QUT), Brisbane, QLD, Australia

Andrew Green AERU, University of Hertfordshire, Hertfordshire, UK

Veeriah Jegatheesan School of Engineering, RMIT University, Melbourne, VIC, Australia

Water: Effective Technologies and Tools (WETT) Research Centre, RMIT University, Melbourne, VIC, Australia

Jaya Kandasamy School of Civil and Environmental Engineering, University of Technology, Sydney, NSW, Australia

Kathy Lewis AERU, University of Hertfordshire, Hatfield, Hertfordshire, UK

Duncan McLuckie School of Civil and Environmental Engineering, University of Technology, Sydney, NSW, Australia

Arturo Aburto Medina School of Science, RMIT University, Melbourne, VIC, Australia

Baden Myers Natural and Built Environments Research Centre, School of Natural and Built Environments, University of South Australia, Mawson Lakes, SA, Australia

David Pezzaniti Natural and Built Environments Research Centre, School of Natural and Built Environments, University of South Australia, Mawson Lakes, SA, Australia

Felicity Roddick School of Engineering, RMIT University, Melbourne, VIC, Australia

Water: Effective Technologies and Tools (WETT) Research Centre, RMIT University, Melbourne, VIC, Australia

Ian D. Rotherham Department of the Natural and Built Environment, Sheffield Hallam University, Sheffield, UK

Kevin Spence Department of the Natural and Built Environment, Sheffield Hallam University, Sheffield, UK

John Tzilivakis AERU, University of Hertfordshire, Hatfield, Hertfordshire, UK

John van Leeuwen Natural and Built Environments Research Centre, School of Natural and Built Environments, University of South Australia, Mawson Lakes, SA, Australia

Future Industries Institute, ITEE, University of South Australia, Adelaide, SA, Australia

Saravanamuthu Vigneswaran School of Civil and Environmental Engineering, University of Technology, Sydney, Australia

Doug Warner AERU, University of Hertfordshire, Hatfield, Hertfordshire, UK

Buddhi Wijesiri School of Civil Engineering and Built Environment, Queensland University of Technology (QUT), Brisbane, QLD, Australia

Chapter 1
Introduction to Urban Stormwater: A Global Perspective

John van Leeuwen, John Awad, Baden Myers, and David Pezzaniti

Abstract Stormwater is seen as a water resource that needs to be captured, stored and utilised in meeting current and future water supply demands. However, increase in impervious surfaces due to expansion of urbanisation has led to increase in pollutants being transported through stormwater runoff and an increased risk of flooding. The world's urban population continues to steadily grow, and the absolute increases in urban populations remain very high and are expected to reach 66% of total world population by 2050. Consequently, ongoing development on managing water resources and water sustainability in urban environments is needed to address risks from increased stormwater flows arising from further development of impervious areas, due to expanded human populations and urban growth. This introductory chapter gives an overview of past and current management of stormwater for flood mitigation, for improved stormwater quality and sustainable practices such as SuDS, LID and WSUD. Existing governance for stormwater management and flood mitigation in selected cities is also included to identify future needs for improving liveability.

Keywords Governance · Harvesting · International · Management · Reuse · Urban · Stormwater

---

J. van Leeuwen (✉)
Natural and Built Environments Research Centre, School of Natural and Built Environments, University of South Australia, Mawson Lakes, SA, Australia

Future Industries Institute, ITEE, University of South Australia, Adelaide, SA, Australia
e-mail: John.vanleeuwen@unisa.edu.au

J. Awad
Natural and Built Environments Research Centre, School of Natural and Built Environments, University of South Australia, Mawson Lakes, SA, Australia

Public Works Department, Faculty of Engineering, Mansoura University, Mansoura, Egypt

B. Myers · D. Pezzaniti
Natural and Built Environments Research Centre, School of Natural and Built Environments, University of South Australia, Mawson Lakes, SA, Australia

V. Jegatheesan et al. (eds.), *Urban Stormwater and Flood Management*, Applied Environmental Science and Engineering for a Sustainable Future, https://doi.org/10.1007/978-3-030-11818-1_1

1.1 Background

Stormwater is an important issue considered by governments at various levels, the building and construction industries, university and other tertiary institutes and the general community. Stormwater has impacts on human living spaces, infrastructure and natural environments and also has value as a water resource. It may be assumed that its meaning is generally understood and agreed on by practitioners and stakeholders. However, definitions of stormwater vary and should be considered in the context of the definition of "runoff" more generally. For example, the Oxford Dictionary defines stormwater as "surface water in abnormal quantity resulting from heavy falls of rain or snow" (Oxford Dictionaries 2017). The National Research Council (2008) in the USA describes stormwater as "that portion of precipitation that does not naturally percolate into the ground or evaporate, but flows via overland flow, interflow, channels, or pipes into a defined surface water channel or a constructed infiltration facility". The Victorian Building Authority (2014) in Australia gives a definition as "rainwater that falls on the ground, paving, driveways or other hard surfaces within a property. It also includes overflows from tanks and roof guttering". Urban stormwater has been defined as "runoff from urban areas, including the major flows during and following rain, as well as dry-weather flows" (Agriculture and Resource Management Council of Australia and New Zealand 1996).

Stormwater generally refers to waters that could impact detrimentally on urban infrastructure and communities including physical damage such as erosion from flows and floods of different scales, increased health risks to communities and ecosystems and transport of chemical and biological pollutants that lower the water qualities of receiving environments (Barbosa et al. 2012; United States Environmental Protection Agency 2012a). It can comprise of materials and constituents accumulated on surfaces including litter, dust and soil, fertilisers, pesticides, micro-organisms, metals, oils and grease (Department of Environment and Heritage 2002) but also constituents less noted for their presence in stormwater, such as per- and poly-fluorinated alkyl substances (Australian Government Department of Defence 2016; City of Salisbury 2016). Through mixing with sewerage effluents in some situations (see Sect. 1.2), higher loads of pathogenic organisms can be collected, in addition to other wastewater-derived pollutants including antibiotic-resistant genes (Garner et al. 2017), endocrine disrupting compounds (EDCs) (Kalmykova et al. 2013) and pharmaceutical compounds (Birch et al. 2015; Zhang et al. 2016).

Increase in impervious surfaces due to expansion of urbanisation leads to the increase in stormwater runoff flow rates, volumes and pollutant loads being transported, which has subsequent impacts on receiving aquatic environments (Aryal et al. 2010; Smith and Fairweather 2016). Increased runoff peak flow and volume physically degrades natural waterways (Walsh et al. 2012). Transport of stream-derived sediment, in addition to pollution from human activity, is a significant source of nearshore pollution including faecal indicator bacteria and viruses (Ahn et al. 2005; Schiff and Bay 2003), nutrients (Fabricius 2005) and suspended solids (Fox et al. 2007). Pollutants from stormwater runoff can also lead to the degradation of a range of coastal habitats i.e. reefs, estuaries and rocky shores (Smith and Fairweather 2016; Cox and Foster 2013; Kinsella and Crowe 2015).

The quality of stormwater in urban and city environments is a function of the contributing catchment characteristics. These include local climate and seasonal variations, land use and degree of imperviousness, existing infrastructure and management, the distance of stormwater from source to receiving environments and current and historic stormwater management practices. Stormwater management is essential for the control of flooding in urban areas and ongoing development is needed to address the impacts from climate change, increasing urban growth and human population expansion.

The world's urban population continues to steadily grow. From 1960 to 2016, the proportion of the global population living in urban areas increased from 33.6% to 52.3% (The World Bank 2017). In China alone, large-scale urbanisation started in the 1980s and increased rapidly from 36.2% in 2000 to 54.8% in 2014 (Li et al. 2017). In Africa, overall urbanisation is expected to increase from about one-third of the population in 2015 to about half the population by 2050 (African Progress Panel 2015). Globally, urban population is expected to reach six billion by 2045 with much of this growth in developing nations. The number of megacities (cities with populations > 10 million) has increased from 10 in 1990 to 28 in 2014, and it can be expected that this expansion will continue (United Nations 2014). Growth in urban population requires urban environments to expand or become more densely populated. In either case, the increased level of built infrastructure and accompanying impervious areas will cause hydrological and ecological changes to receiving waters (Burns et al. 2012) if stormwater management is undertaken in the conventional manner of rapid disposal, as described in Sect. 1.2.

1.2 Past and Present Stormwater Management Practice

The management of stormwater has historically been driven by the need to protect the community from the adverse (sanitary and flood) impacts created by rainfall and runoff. This has led to the development of hydraulically efficient systems for the collection and conveyance of runoff to a receiving location, typically a stream, lake or ocean outfall. Across many jurisdictions, the primary objective for stormwater management that continues to exist is to provide protection to the community. Authorities at several levels are charged with ensuring a standard of flood protection is provided to the public. In many instances, these authorities have developed guidelines and standards for setting minimum requirements for flood management, and multiple guidelines can exist even in a single country. For example, in Australia, there are guidelines at the national level, namely the Australian Rainfall and Runoff

Guidelines (Ball et al. 2016) and specific to some state governments such as Queensland (Institute of Public Works Engineering Australia 2013). For flood protection, these guidelines use the magnitude and frequency of runoff events as a means to provide a level of protection. Typically, there are two standards of flood protection proposed by these guidelines (Ball et al. 2016):

1. Higher-frequency, lower-magnitude runoff events – events expected to occur with a shorter average recurrence interval, e.g. 2–10 years – to enable safe pedestrian and vehicle passage in urban areas
2. Lower-frequency, higher-magnitude events – events expected to occur with a longer average recurrence interval, e.g. greater than 50 years – that should be managed to protect life and infrastructure

In Australia, stormwater runoff has been managed using a network of pipework systems separate from sewer systems (Department of Environment and Heritage 2002), while in other countries such as the USA (Southeast Michigan Council of Governments 2008), China (Che et al. 2011; Wu et al. 2016; Li et al. 2014), the UK (Southern Water, accessed 2017; Dwr Cymur Welsh Water, accessed 2017) and some European countries (Ashley et al. 2007), there are many cities where stormwater management includes the use of combined sewer systems where stormwater and wastewater are conveyed together via a single pipe drainage system.

1.2.1 Drainage System Approaches

Mixed storm and sewer waters and combined sewer overflow can present a wide range of risks from contaminants such as EDCs, pharmaceuticals, personal care compounds (Ryu et al. 2014) and pesticides (Park et al. 2017). Combined sewer waters present challenges in terms of treatment requirements, especially during high flow events. Management for this includes temporary storage of excessive water flows (e.g. Wu et al. 2016) until these can be treated at a wastewater treatment plant or diverted without treatment to receiving environments (e.g. Southeast Michigan Council of Governments 2008).

Due to the benefits of separation, recent and ongoing infrastructure developments globally are generally implementing separate stormwater and sewer systems (Southeast Michigan Council of Governments 2008). However, despite the separation of stormwater from sewer systems, wet weather conditions can still lead to overflows from sewerage systems, which impact on waterways and cause significant problems (Department of Environment and Energy 2002).

According to one Australian source (Victorian Builders Authority 2014), stormwater can pose the same risks as sewage effluent and should always be treated before any reuse. Li et al. (2014) reported a study on the performances of separate sewer systems (SSS) and combined sewer systems (CSS) of Shanghai and Hefei, finding that serious illicit connections occurred for some SSS. Their results showed that for the systems investigated, there was no obvious advantage of having SSS over CSS in terms of pollutant control, suggesting that the mix of stormwater and sewer waters presented the same risks as sewer wastewaters alone. Park et al. (2017) reported that many studies have focussed on organic wastewater contaminants such as EDCs and pharmaceuticals and personal care products (PPCPs), with the underlying assumption that the chemicals of concern come from point source municipal wastewater treatment facilities. However, they reported that significant EDCs and PPCPs may also come from nonpoint sources such as agriculture and urban/suburban runoff and hence are potential contaminants in stormwater.

Despite the infrastructural separation of this water resource from sewerage networks, significant challenges still exist in this form of water supply. These include high variability in flows from extreme rainfall events, prolonged droughts and costly infrastructure for stormwater utilisation that includes capture, treatment (for traditionally recognised pollutants) and separate pipelines for distribution to users. There are further risks from pollutants derived from catchment-specific human activities that have gone unrecognised until these are subsequently detected in downstream stormwater, as demonstrated by contamination by Per- and poly-fluoroalkyl substances (PFAS) derived from past use as a fire retardant. For example, at a military defence base in South Australia (Australian Government Department of Defence 2016) within a stormwater catchment of the City of Salisbury, low levels of PFAS have been detected in drains, wetlands and aquifers near to that base (City of Salisbury 2016), which subsequently affect stormwater harvesting operations.

1.2.2 New Approaches to Stormwater Management

In Australia, with much traditional stormwater management being focussed on rapid transport from urban areas to receiving waters, stormwater had received little, if any, treatment (Department of Environment and Energy 2002). In many instances, the rapid collection and transport of runoff has led to flooding in downstream reaches, requiring peak flow mitigation typically in the form of runoff volume detention, where stormwater is temporarily stored and released at a controlled flow rate without treatment (Argue and Pezzaniti 2005). Other practices for ameliorating flooding involve retention practices where runoff is captured and either allowed to infiltrate into soil on site or be used as an alternative water resource. In some instances, stormwater systems can provide multiple benefits for flood control through stormwater detention and retention.

Historically, the management of stormwater has focussed attention at the point of discharge. However, it is now recognised that better outcomes (e.g. stormwater quality) can be achieved if efforts are directed throughout the catchment to minimise treatment needs (Environment Protection Authority South Australia 2007). This emulates a Hazard Analysis and Critical Control Point (HACCP) approach that is well established in the food and beverage industries (Bradsher et al. 2015) and for management of drinking water quality and supply in Australia (Australian Drinking Water Guidelines 2011).

According to the Australian Department of the Environment and Energy (2002), a well-designed, integrated stormwater system can provide a range of community benefits including minimisation of flood risk, protection of downstream water bodies, preserving aesthetic values, recreational facilities, natural habitat conservation and water reuse. The Department of Environment and Energy (2002) details that this can be achieved by management of source control, contaminant interception and receiving waters.

More recent stormwater management approaches globally include enhanced stormwater retention and rainwater infiltration as part of programmes such as water-sensitive urban design (WSUD) in Australia, the "Sponge City" initiative in China (Li et al. 2017; Austrade 2016), sustainable urban drainage systems (SuDS) in the UK (NetRegs, accessed 2017) and low impact development (LID) in the USA (United States Environmental Protection Agency 2012b). Dhakal and Chevalier (2016) referred to such approaches alternatively as green infrastructure (GI). The background and definition of these and other similar approaches has been well documented by Fletcher et al. (2015). For these, the focus is on managing water resources in urban environments that includes addressing risks from increased stormwater flows arising from further development of impervious areas from expanded human populations and urban growth (Li et al. 2017; United States Environmental Protection Agency 2012b; Chandana et al. 2010). Stormwater management strategies are formulated on the basis of increasingly recognising the need to emulate predevelopment flow regimes in addition to reducing pollutants and their loadings (Hamel and Fletcher 2013).

Current advances in stormwater management such as LID, WSUD, SuDS and Sponge Cities involve a shift in approach from a reactive response through engineered actions to more interactive and collaborative water management styles (Yang 2016). For instance, in the Netherlands, the "Space for Rivers, 2007" and the "Building with Nature" programmes under the Climate-Proof City initiative involve taking approaches to stormwater management as working or building within the scope of naturally occurring variances, i.e. "a paradigm shift from building in nature to building with nature" (Yang 2016; de Vriend and van Koningsveld 2012).

1.2.2.1 Water-Sensitive Urban Design

The ongoing and increasing application of WSUD principles in Australia has enabled the control and better management of stormwater runoff, minimisation of pollution and for storage and reuse to occur in urban, commercial and industrial settings (Department of Environment and Energy 2002). The peak professional body in Australia (Engineers Australia) for stormwater management has recognised the need to provide guidance on broader management of stormwater, including integrated water management approaches such as WSUD (e.g. Wong 2006 and Ball et al. 2016). Local jurisdictions have also taken a broader approach to managing stormwater. For example, the Australian Capital Territory urban infrastructure and stormwater design standard (ACT 2009) refer to their water policy plan, which lists a range of objectives, and design policies/standards to:

- provide safety for the public
- minimise and control nuisance flooding and to provide for the safe passage of less frequent flood events
- stabilise the landform and control erosion
- protect property from flooding
- enhance the urban landscape
- optimise the land available for urbanisation,
- minimise the environmental impact of urban runoff on water quality
- provide opportunities to enhance the environment through the provision of water sensitive stormwater design.

Direct rainwater harvesting from roofs to storage tanks for potable use as part of WSUD is commonly used to supplement potable water supply in some Australian states for uses including toilet flushing, washing, bathing, laundry and gardening (Government of South Australia 2010; Australian Bureau of Statistics 2013). Some uses of harvested rainwater then mitigate the volumes of stormwater that otherwise would flow to drainage systems. Built infrastructure for stormwater transport, treatment, infiltration, storage and reuse includes use of swales, detention and sedimentation basins, constructed wetlands, permeable pavements with underground storage and managed aquifer recharge and reuse (Government of South Australia 2010). The high variance in the quality and quantity of stormwater leads to high variance in human perspectives on its value as a resource and the risks it poses on built and natural environments.

Under the WSUD framework, stormwater is seen as contributing to all of the available water resources that include rainwater harvesting, stormwater, reclaimed wastewater (industry and domestic), potable water and other sources such as local bore water (Government of South Australia 2017). Practices under WSUD are to promote the sustainable use and reuse of water in urban environments and can be applied to residential, council, commercial and industrial developments, buildings and amenities. According to the Government of South Australia (2010), WSUD helps communities to achieve greater water sustainability and provide pleasant places to live and work. This involves a wide range of practices relating to the supply, sustainability and management of water resources for urban environments. These include integrated management of water resources (e.g. stormwater, groundwater, potable water and wastewaters) for protection of water-related environmental, recreational and cultural values. Management involves storage, treatment and reuse of stormwater and wastewater, use of vegetation for treatment purposes and utilisation of water-saving measures.

WSUD has a key focus on the management of stormwater for its capture and reuse, most especially in dry climate regions and/or when climate conditions shift to prominent dry conditions during El Niño events. For example, a cyclic change in focus from the management of stormwater to mitigate flood risk to an emphasis on its use as a highly valuable water resource occurs in Adelaide and other urban areas of South Australia. South Australia is noted as being the driest state of the driest inhabited continent of the world. The urban areas of Adelaide have mean annual rainfalls from approximately 400–600 mm (with an annual average pan evaporation of 1600–1800 mm) and, like most of the Australian continent, are highly influenced by the Walker circulation. As such, extreme variation in climatic conditions can occur with varying strengths of El Niño and La Niña events. Under strong El Niño events, ongoing drought conditions prevail, and available surface waters can be limited, restricting water supplies and providing impetus to WSUD development for stormwater capture and reuse. In contrast, under strong La Niña events such as those that occurred in 2010/2011 and 2011/2012, with subsequent very wet conditions, emphasis shifted to the management of stormwater flows to mitigate flood risks. In this case, the capture and reuse of stormwater remains, but under the latter climate conditions, dependency on this for water supply is diminished.

1.2.2.2 Sponge Cities

Sponge City refers to "sustainable urban development with enhanced flood control, water conservation, water quality improvement measures and natural eco-system protection" (Li et al. 2017). Guidelines for the Sponge City programme were issued in 2014, and in 2015, 16 cities in China were chosen for a 3-year implementation programme (Austrade 2016). The Sponge City programme has been described as a multifunctional and cross-sectoral solution for stormwater management (Yang 2016). Under this programme, a city's infrastructure and features are managed as "a sponge" to improve water quality, for storage and for reuse purposes (Li et al. 2017). The main objectives for China's Sponge City programme include retaining 70–90% of the average annual rainwater onsite through implementation of GI, preventing urban flooding, improving urban water quality, mitigating impacts on natural ecosystems and alleviating urban heat island impacts (Li et al. 2017; Yang 2016). The scope of each project includes implementation of "Sponge City" management actions at neighbourhood scale including parks, greenbelt, highways, water system restoration and sewage recycling (Beijing Capital Group Company 2017). Implementation of the programme involves application to 20 square kilometres in demonstration cities (Yang 2016). A further objective of the programme is to strengthen urban planning and construction management (Austrade 2016). According to Yang (2016), the programme is designed to address the main causes of urban flood in China, which include climate change, urbanisation, inadequate urban planning and lagging infrastructure development.

1.2.2.3 Sustainable Urban Drainage Systems

In the UK, current stormwater management practices are under a regulatory framework and are referred to as "sustainable urban drainage systems" (SuDS). These are designed as a natural approach for managing drainage in urban areas and to minimise flooding and water pollution. These are also designed to generate green spaces and habitats for fauna and flora in urban environments (NetRegs, accessed 2017). SuDS are a legal requirement for all new developments except for surface water drainage from single dwellings and developments that drain to coastal waters (NetRegs, accessed 2017).

SuDS function to slow the flow and to retain water runoff from a site to allow natural processes to degrade pollutants. Under SuDS, it is recommended that flows between SuDS components are connected using swales, filter drains or ditches and avoid the use of pipes. Through this approach, more stages of treatment are implemented to remove contaminants from stormwater. Overflows from site controls are treated at regional controls before being discharged to the environment. SuDS technologies include green roofs, permeable surfaces, infiltration trenches, filter drains and filter strips, swales, detention basins, purpose-built ponds and wetlands (NetRegs, accessed 2017).

1.2.2.4 Low Impact Development

Similar approaches to stormwater management are implemented in the USA and Canada. According to the United States Environmental Protection Agency (2012a), low impact development (LID) is designed to reduce runoff and pollutant loadings by managing runoff close to its source(s). LID includes overall site design and management and individual, small-scale stormwater management practices. These are based on the use of natural systems for infiltration, evapotranspiration and use of captured rainwater. The US EPA considers that holistic approaches maximise the benefits of LID. Benefits of LID have been listed by the US EPA as follows: improved water quality, reduction in flood events, restoration of aquatic habitats, improved groundwater recharge and enhanced environments (United States Environmental Protection Agency 2012b). Other benefits reported include mitigating heat island effect, energy saving, reduced air pollution (through energy savings in GI developments) and increased property values (improved aesthetic living environments).

1.2.3 Summary of Approaches

Both SuDS and LID approaches have an emphasis on stormwater management in terms of reducing risk from urban floods, managing stormwater to minimise its pollution and lessening risks to combined stormwater and sewer outflows. Apparent further benefits include groundwater recharge and enhanced urban environments for both human and wildlife well-being. Under the Sponge City programme, as with WSUD, the consideration of reclaimed domestic wastewaters as a further water resource is considered, and WSUD also considers water demand management.

1.3 Water Resources and Stormwater Harvesting

In Australia, increase in water demand and security of supply associated with population growth and limitation in surface water resources under dry climate conditions have forced federal, state and local governments and the water industry sector to seek alternative water resources. This includes the desalination of seawater, domestic wastewater recycling and harvesting and use of urban stormwaters. The reuse of domestic wastewaters has faced greater public resistance and scrutiny in comparison to the capture and use of stormwater (Coombes et al. 2006), but now the use of tertiary-treated domestic wastewaters for non-potable uses such as for gardens and toilet flushing, and particularly for horticulture, has become widely accepted.

In recent years, the management and use of stormwater has become part of the total water resource supply and management. The availability and security of each supply source is a result of historic occurrences (including extreme climate conditions) and subsequent government decision-making and actions taken to address current and future risks to water resources and to meet existing and projected future supply demands.

From late 1996 to mid-2010, most of southern Australia (except parts of central Western Australia) experienced a prolonged drought known as the Millennium Drought. This drought affected the Murray-Darling Basin, the major catchment for water resources for southern Queensland, New South Wales, Victoria and South Australia (Bureau of Meteorology 2015).

During this period, water resources for agriculture, domestic use and environmental and ecological health became increasingly limited and restricted in eastern and southern states. This heightened the general awareness and concern around the limitations and insecurity in water supply to meet community demands. Under these conditions, state and federal government actions included investment in development of large-scale seawater desalination plants in all mainland states for potable water supply. Nonetheless, Australia's mainland experienced above average rainfall due to heavy monsoonal rainfall in the north, illustrating the high variability in water sources geographically, seasonally and over short- and long-term periods. The development of the desalination plants was designed to provide security in potable water supply, i.e. to "drought-proof supply", and enable alleviation of previous severe water restrictions which had been progressively implemented by government and water authorities. With long-term droughts and increases in water supply demand due to a growing population and urbanisation, water resource management in western, southern and eastern Australian states and regions also included domestic wastewater reuse schemes at council/township level and larger-scale operations. In dry and arid regions, such as those of Western Australia, there is extensive use of reclaimed water for watering of sports ovals and parklands (Western Australia Water Corporation 2005). Large-scale domestic wastewater reuse schemes have also been developed to meet water supply demands for intensive horticultural practices. These include the Werribee Irrigation District Recycled Water Scheme in Victoria, which was designed to address water shortages and to secure water supply for horticulture.

In South Australia, wastewater reuse schemes include the Virginia Pipeline Scheme (VPS) that supplies reclaimed water to horticultural practices, the Glenelg-Adelaide Parklands Recycled Water Project and the Southern Urban Reuse Scheme that provides reclaimed water for ~8000 homes. Being generally of low salinity, stormwater can be blended with brackish waters to improve water quality. For example, stormwater has been blended with tertiary-treated brackish recycled domestic wastewater for a second reticulated water supply to a community of 10,000 residents in South Australia (Page et al. 2014; Hains 2009; Department of Planning Transport and Infrastructure 2013). The South Australian Government (Department of Planning Transport and Infrastructure 2013) listed 19 key reclaimed water schemes involving the reuse of domestic wastewater and stormwater that are operational in South Australia.

The reuse of stormwater and wastewater is considered important in achieving sustainable urban water supplies (Ferguson et al. 2013; Furlong et al. 2017). Benefits from stormwater reuse include the reduction in the transport of pollutants to receiving environments including streams, rivers and coastal waters. These pollutants can include sediments from upstream sources and urban environments including construction sites, roads, pavements and car washing, leading to reduced light penetration in aquatic environments, limiting aquatic plant growth and benthic organisms. Nutrients sourced from detergents, spillages, fertiliser use and animal/bird faeces can lead to excessive growth of algae, cyanobacteria and aquatic weeds, as well as reducing dissolved oxygen levels in aquatic environments (Victorian Stormwater Committee 1999). Degradation of coastal environments includes losses of seagrasses from excess nutrients (nitrogen) and sediments. From the results of a major study on the coastal environment in Adelaide, South Australia, a key recommendation was that stormwater and wastewater flows be reduced (Fox et al. 2007). Furlong et al. (2017) investigated the impact of two stormwater reuse projects in southeastern Australia and reported reductions in nitrogen discharged to local watercourses of ~ 3–4 kg/ML.

Aquifer storage and recovery (ASR) is a stormwater harvesting approach that can be used strategically for the storage of stormwaters for recovery when needed (Pyne 1995). It is used as part of the treatment process to improve and attain fit-for-purpose water quality. A further advancement in the practice is aquifer storage, transfer and recovery (ASTR), where extraction is from dedicated bores spatially located away from injection bores to ensure transport of waters in aquifer media over determined time periods (Page et al. 2014). A well-documented example is the Parafield ASR system within a catchment (1590 Ha with 73% urban area) of South Australia that comprises use of a retention basin, constructed wetlands and an ASR for harvesting and treatment of stormwater. Recovered waters are used for industries, parklands and ovals and residential non-potable uses. The water quality of wetland outflow waters and waters recovered from the aquifer, based on physical and chemical parameters, has been reported to meet drinking water quality criteria except for iron, turbidity and colour, which occasionally exceed the Australian Drinking Water Guidelines (Dillon et al. 2014). However, although micro-organic pollutants such as herbicides (e.g. diuron, simazine, atrazine, metolachlor and chlorpyrifos), polyaromatic hydrocarbons (PAHs, e.g. fluorene) and the flame retardant, Tris(1-chloro-2-propyl)phosphate, have been detected (at ng/L levels) in pre- and post-ASR/ASTR stored stormwaters, the efficiencies of attenuation have been difficult to determine (Page et al. 2014). From a further recent study by Page et al. (2017), it was found that ASR of stormwater led to significant decreases in *E. coli* levels at four test sites in South Australia, but for metals/metalloids and nutrients, the trend was less clear.

The capture and storage of stormwater for large-scale reuse purposes presents clear benefits but also continued challenges. All schemes require significant investment in infrastructure, including pumps and distribution networks. Options such as ASR and ASTR schemes require suitable locations from where waters can be distributed and used to meet demands. Groundwater quality and hydrology also need to be understood before selecting sites to install bores. Alternatively, open surface storage stormwater harvest schemes, such as constructed wetlands, dams and retention basins, present challenges including land availability, new forms of contamination (e.g. cyanobacteria and toxins), concentration of existing water contaminants and losses from evaporation.

In some cases, stormwater is a natural water resource, ultimately for domestic and potable supply. The City of Mt. Gambier in South Australia is an example of a catchment with an undulating topography with no streams. Stormwater is managed by directing storm flow into bores, resulting in direct recharge of an aquifer that then flows into an extinct crater (the Blue Lake) that is used for domestic water supply. Management guidelines (Environment Protection Authority South Australia 2007) include that recharge to the aquifer should be by infiltration rather than direct recharge into bores in order to satisfy regulatory water quality criteria. Risks to water quality are naturally attenuated through chemical adsorption processes and physical filtration as waters move slowly though the limestone aquifer.

1.4 Stormwater Treatment Measures

Urban stormwater is increasingly perceived as a valuable water resource (with treatment and storage) for various applications. In Australia, this includes agricultural use, irrigation of parks and gardens, industrial and commercial uses, GI, non-potable uses such as toilet flushing and washing and potable use (Gerrity et al. 2013). Benefits from stormwater use include reduction in the demand of potable waters. For example, in the Adelaide and Mount Lofty Ranges, South Australia, it has been estimated that ~ 1 GL/y of water can be saved from traditional water resources by the use of stormwaters from schemes involving treatment by wetlands followed by ASR (Government of South Australia 2010).

Design and selection of treatment trains depend on stormwater quality and target outflow quality that minimise environmental and public health risks and meet end-use objectives. Treatment measures include sand filtration, biofiltration, gross pollutant traps, grassed swales, sedimentation ponds and wetlands (Aryal et al. 2010;

Hatt et al. 2006). Constructed wetlands, ponds and sand filters can be used to reduce pathogen levels, while physical (UV irradiation) or chemical (chlorine, ozone or iodine) disinfectants can be used to kill/inactivate pathogens found in stormwater (Department of Environment and Conservation 2006). Nolde (2007) reported that in Berlin-Lankwitz, Germany, stormwater collected from roofs, streets and courtyard surfaces and stored in rainwater reservoirs (190 m^3) is treated by biological filtration followed by physical disinfection (UV irradiation), providing water for toilet flushing and garden watering. Stormwater treatment measures are considered in more detail in Chaps. 4 and 5.

Constructed wetlands provide for natural processes to reduce pollutants, including nutrients from stormwaters. Adyel et al. (2016) reported percentage reductions in total nitrogen and total phosphorus levels at 62–99%, respectively, during dry weather, and at 76–68% during wet weather conditions. Constructed floating wetland systems are a novel stormwater treatment technology currently being trialled in Australia (Schwammberger et al. 2017). In a study conducted on a constructed floating vegetation wetland (planted with *Carex appressa*) in South East Queensland, Walker et al. (2017) reported percentage reductions in total suspended solids and total phosphorus levels of 81–52%, respectively. In contrast, the percentage reduction in total nitrogen was only 17%.

Pervious pavement is another method that is used to control pollutant levels in stormwater, by filtration processes. According to Mullaney and Lucke (2014), the four common types of pervious pavements (i.e. concrete and plastic grid pavers, permeable interlocking concrete pavers, porous asphalt and porous concrete) have high infiltration rates and high efficiency in removal of various stormwater pollutants, i.e. suspended solids (e.g. ~64%, (Legret et al. 1996)) and heavy metals (e.g. over 94% for zinc, cadmium and lead (Dierkes et al. 2005; Myers et al. 2011)). Nnadi et al. (2015) reported that the quality of harvested stormwater from a pervious pavement system had chemical quality that met various international agricultural irrigation standards. In a study by Jayasuriya and Kadurupokune (2010), the filtration and detention properties of a pervious pavement showed potential to mitigate floods by reducing the peak discharge of stormwater by 43–55%.

Trowsdale and Simcock (2011) reported a multilayered bioretention system (comprising a drainage layer of coarse sand, subsoil 600–700 mm, topsoil 300–400 mm and composted mulch, 80 mm) reduced the concentration of suspended solids (medium percentage removal, 90%), zinc (95%) and lead (91%), while the copper level increased (+50%) as the system released copper. In contrast, Davis et al. (2003) and Hatt et al. (2007) reported that more than 90% of total zinc, lead and copper were removed for all studied filter media (six different soil mixtures). The differences in removal efficiencies were attributed to the differences in the bioretention systems, i.e. their filter media (Roy-Poirier et al. 2010). Bioretention systems are effective in the control of nutrients in stormwaters with the removal efficiency dependent on plant species used and the organic content in the soil mixture (Roy-Poirier et al. 2010; Bratieres et al. 2008).

Successful implementation of stormwater technologies require an integrative approach, where a combination of methods are utilised and a single technology is unlikely to solve issues concerning excessive runoff and flooding (de la Trincheria and Yemaneh 2016).

1.5 Stormwater Governance and Management

According to Olsson and Head (2015) and further reported by Chattopadhyay and Harilal (2017), essential roles of urban water governance include the management of environmental dynamics to provide reliable water supplies for cities.

Forms of governance in water management include hierarchical, market and network-negotiated (Porse 2013; Brown et al. 2011; Kjaer 2004; Pierre and Peters 2000). Where cities and urbanised regions have well-established stormwater systems, governance is often of a well-developed, hierarchical structure (local, regional/state, national) that includes government, industry, private and community organisations (Porse 2013). In many cases, national agencies establish guidelines for stormwater management and provide some funding (Porse 2013), e.g. for new stormwater management programmes and initiatives that occur in Australia (Environment and Communications References Committee 2015). State and provincial (district or local government) authorities provide regulations and administrative support for stormwater management. At the local level for established systems, stormwater management responsibilities have traditionally been undertaken by municipal governments or municipal agencies, with other agencies such as those dealing with transport or recreation, also having some responsibilities (Porse 2013). Cities without major stormwater infrastructure typically lack effective governance structures. For these cities, stormwater management may rely on resident and community group networks with city planners. In such cases, and where municipally managed systems exist, they are often operated with less regulation and with national and regional funding and expertise and are more prone to corruption (Porse 2013).

Stormwater governance, in the context of traditional urban stormwater management, has focussed on rapid drainage and removal of stormwater through centralised systems and as such is structured to support these conventional systems (Dhakal and Chevalier 2016). Although stormwater infrastructure has traditionally been designed for conveyance, there is increasing GI development in cities that integrate conveyance and infiltration, in hybrid systems (Porse 2013) (see Sect. 1.2). Such hybrid systems involve distribution of management responsibilities including for planning, operations and maintenance (Porse 2013). Effective governance of stormwater management responsibilities is required that accounts for the diversity and interactions of infrastructures comprising these hybrid systems. According to Porse (2013), this requires relevant expertise to exist in agencies and other stakeholders such as businesses and private landholders. In hybridised governance, management and monetary responsibilities are shared between them. According to Dhakal and Chevalier (2016), centralised governance is unsuitable for systems that need a distributed management approach, where multiple stakeholders are involved. Zhang et al. (2016) compared conventional and decentralised stormwater management for Singapore and Berlin and based on cost analyses considered that a combination of both management approaches would likely be the most practical for most cities.

In relation to governance of urban water in both developing and developed nations, Olsson and Head (2015) detailed key challenges as follows: competing interests among different sectors/stakeholders, effective involvement of citizens and stakeholders and different interpretations of integrated water management. Some aspects of current governance and stormwater management issues for several regions and countries are briefly detailed below.

1.5.1 Australia

In Australia, the management of stormwater is the responsibility of state and local governments, with the federal government being involved in and supporting national stormwater management and development initiatives. A summary of stormwater governance is provided by a recent parliamentary inquiry into stormwater management (Environment and Communications References Committee 2015). In recent years, Australian Government initiatives for stormwater capture and use have included programmes to encourage innovation in sourcing, treatment, storage and discharge. These initiatives have also been to encourage agreed actions by stakeholders to promote innovation and capacity building for establishment of water-sensitive Australian cities (Environment and Communications References Committee 2015). Under recently completed programmes such as the *National Water Security Plan for Cities and Towns*, *Water Smart Australia* and *Strengthening Basin Communities*, stormwater capture and reuse projects are federally funded with an objective to replace and conserve potable water supplies.

The Australian Government, through its Department of Environment and Energy, has provided information on urban stormwater management for the Australian public (Department of Environment and Energy 2002). The Commonwealth of Australia also supports *Australian Rainfall and Runoff*, which is a national guideline for stormwater and drainage management (Ball et al. 2016). This information is provided with no legal liability being accepted and is advisory only. Further to this, a National Water Quality Management Strategy (NWQMS) policy objective is "to achieve sustainable use of the nation's water resources by protecting and enhancing their qualities while maintaining economic and social development". Under the NWQMS is the Australian Guidelines for Urban Stormwater Management (Department of Environment and Energy 2002).

Australian Government action on mitigating stormwater impacts includes provision of funds for programmes to demonstrate ways to improve coastal and marine water quality. The funding is targeted for councils of coastal regions, state government agencies, industries and water management organisations with the aim to promote best practice and innovation. This is to achieve substantial beneficial impact on water quality in coastal areas and cities (Department of Environment and Energy 2002).

Urban stormwater is generally managed by local government (councils) with state and territory governments having overall responsibility for land and water use planning and management (Department of Environment and Energy 2002). Councils of cities and metropolitan areas typically develop a stormwater management plan (SMP) that includes capital works and services, programmes for asset management, protecting the environment and promoting ecological sustainability (Department of Environment and Energy 2002). By 2015, efforts had been made to incorporate WSUD principles into planning and development processes at both state/territory and local government levels (Cook et al. 2015). Currently, Victoria and the Australian Capital Territory (ACT) only have some mandated requirements for WSUD targets for Greenfield developments. In other jurisdictions, state-level policies, targets and guidelines provide a framework, and implementation is at the local government level (Cook et al. 2015). At the local government level, the considerations of WSUD are both water quality and quantity of waters. Tjandraatmadja et al. (2014) researched WSUD design impediments and potential in South Australia and reviewed WSUD legislation and policies across Australian states and territories.

1.5.2 Baltic Sea Region

In the Baltic Sea Region, policies and legislation for urban stormwater management include the following: the Water Framework Directive 2000/60/EC (WFD), Floods Directive 2007/60/EC, Urban Wastewater Directive (91/271/EC) and Environmental Quality Standards Directive (2008/105/EC). Policies and regulations specific to Germany, Latvia, Finland, Sweden and Estonia are also detailed through an EC Project Report – Baltic Flows (de la Trincheria and Yemaneh 2016).

The management of water resources in Germany is by three levels of authority, being at federal, state and municipal (local government) levels. These levels have been described as a well-developed hierarchical structure by Porse (2013). Germany was one of the first countries to implement rainwater and stormwater management measures into policies based on impervious surface cover and incentivised through tax reductions (de la Trincheria and Yemaneh 2016). Sweden also has a hierarchical (top-down) structure for water resources management (de la Trincheria and Yemaneh 2016). In Latvia, rain (storm) water management is regulated by national legislation and is enforced through local (municipal) government, mostly by control of infrastructure development. In Latvia, the maintenance of stormwater infrastructure is mainly the responsibility of municipal authorities or water companies. Stormwater conveyance infrastructure design is according to Latvian Construction Norm LBN 223–99 based on maximum calculated runoff and has not considered rainwater retention or infiltration (de la Trincheria and Yemaneh 2016).

1.5.3 Canada and the USA

Dhakal and Chevalier (2016) described existing stormwater governance in the USA as centralised, hierarchical and structured with support from federal, state and local governments. The US EPA enforces federal laws through standards and regulations, and states can enact and implement standards on city and county governments. City and county governments implement federal and state laws and can enforce their own discretionary standards and regulations where minimum standards must comply with the state and federal standards (Dhakal and Chevalier 2016). Although GI has been encouraged in many US cities for over a decade, it is not widespread in application, detailing existing governance as a major barrier to mainstreaming GI (Dhakal and Chevalier 2016). This includes the inability of cities to enforce National Pollutant Discharge Elimination System (NPDES) permits (under the Clean Water Act) for discharge of pollutants through stormwaters from private lands. This is because stormwater is not defined as a pollutant and private land is not considered as a point source. Further, the Fifth and Fourteenth amendments to the US constitution prohibit government from the taking or controlling of private lands without just compensation, which is ill defined or involves complex legislation (Dhakal and Chevalier 2016), which could make stormwater runoff controls difficult to implement. Despite this, some US cities, including Chicago, Seattle, New York and Washington, have been implementing GI with encouraging results (Dhakal and Chevalier 2016).

In Canada, various legislations exist to support urban water management, including stormwater management. For example, the Province of Ontario's Development Charges Act (1997) enables municipalities to levy for off-site costs associated with infrastructure developments. Funding acquired by municipalities by the above legislation provides support for water system upgrades and their expansion in the Ontario province. Under the Ontario Water Resources Act (1990), water quality and quantities of surface and groundwater resources are protected. Through this Act, the discharge from municipal wastewater treatment plants and stormwater management is regulated (IANAS 2015). Toronto's municipal government is responsible for stormwater management. Funding is acquired through development charges that support infrastructure capital projects and thereby future urban growth (IANAS 2015).

1.5.4 Latin America

Urban flooding is a major issue in Latin America where infrastructure planning is insufficient and where drainage and sewerage system developments are not keeping pace with rapid urban growth (Reuters 2017). According to Nalesso (reported by Reuters (2017)), the issue of disaster risk prevention was not a priority for many governments in Latin America (in 2017).

Urban water supply, service and management issues and recommended ways to resolve the significant water resource problems of Latin American and Caribbean countries have been detailed in an InterAmerican Network of Academies of Sciences (IANAS) report, in 2015 (IANAS 2015). For these countries, urban water challenges are broad ranging. They include the provision of potable water to remote communities and to economically poor communities such as those living in urban environments outside of planned living zones and within floodplains. Limitations in wastewater collection services and low percentage levels of wastewater treatment also lead to significant water pollution through discharge to local freshwater bodies such as rivers. As a result, those dependent on these resources may experience high frequencies of waterborne diseases (including diarrhoea, amoebiasis, malaria and dengue) and severe floods. Other noted problems include limitations in the following: resources (monetary, human capital – skills and knowledge – infrastructure), planning, integration of water-related servicing authorities and organisations, jurisdictions not matching catchment basin-scale management needs and corruption within responsible authorities and organisations. Deforestation has been noted as a cause of low water quality. Regular and major floods and their risk management are a significant challenge for countries such as Argentina, Uruguay, Costa Rica and Venezuela. Aspects of urban water management and governance, some relating to floods and control, are given as follows. For the City of Buenos Aires, Argentina, a water management plan had been implemented that identified the key causes and impacts of floods and set work guidelines to mitigate impacts from frequent floods (Lopardo et al., *in IANAS (*2015*)*). According to Tundisi et al. in *IANAS (*2015*)*, Brazil has current legislation and technologies for protection of water sources and its distribution, though these have not led to improvements in source water quality. These authors further recommended the incorporation of reuse of treated wastewaters in water management plans in addressing critical water shortage needs and to include the concept of green cities in all programmes for integrated water management of urban regions. In Brazil, at the municipality level, there are specific laws for regulating urban development. The integration of federal, state and municipal legislation is a key challenge in providing water supply (Tundisi et al. in *IANAS (*2015*)*). In Bolivia, the federal government establishes standards, while local and regional governments are responsible for solving problems inside their jurisdictions. Hydrographic basin management is not aligned with the administrative jurisdictions, which complicates the management of water resources (Urquidi-Barrau in *IANAS (*2015*)*). Significant challenges include political and economic barriers to establish and implement new urban water management legislation for integration and coordination of plans and/or projects (Urquidi-Barrau, in *IANAS (*2015*)*).

For Costa Rica, urban flooding has been reported to be related to three key factors: inadequate capacity of stormwater conveyance infrastructure, land use changes and climate change (with increase in extreme events) (Hidalgo León et al., in *IANAS (IANAS* 2015*))*. Flood management and control is generally through conveyance (drainage) systems. In Costa Rica, the Servicio Nacional de Aguas Subterráneas, Riego y Avenamiento (National Service of Groundwater, Irrigation and Drainage, SENARA) deals with irrigation, drainage and flood protection. Objectives established for SENARA are in Law No. 6877, which define its strategic roles (Hidalgo León et al. in *IANAS (*2015*)*).

In Uruguay, over 60 urban centres are affected by stormwater drainage problems with 70% classified as moderate to serious (Capandeguy et al. in *IANAS (*2015*)*). For Uruguay, floods are the main factor activating the National Emergency System, and a challenge is to improve planning systems and incorporate more sustainable infrastructure. Better information and information access are noted as being needed to achieve sustainable urban water management. Stormwater has been identified as potentially being able to provide benefits for cities of Uruguay. However, its management has focussed on conflict resolution, which conceals its potential (Capandeguy et al. *in IANAS (*2015*)*).

1.5.5 Republic of South Africa (RSA)

The RSA has enacted the National Water Act (NWA, Act No 36) (Republic of South Africa 1998) for water management, and its purpose is to protect, use, manage and control the country's water resources and to establish institutions that aim to achieve this purpose. The NWA (Republic of South Africa 1998) recognises the need to protect the quality of water resources and the need for integrated management of all aspects of water resources. The NWA (Republic of South Africa 1998) allows households to harvest rainwater (i.e. Schedule 1(c); "A person may, subject to this Act – store and use run-off water from a roof") and points out that stormwater should be treated prior to discharge (i.e. Schedule 1(f); "A person may, subject to this Act – discharge run-off water, including stormwater from any residential, recreational, commercial or industrial site, into a canal, sea outfall or other conduit controlled by another person authorized to undertake the purification, treatment or disposal of waste or water containing waste, subject to the approval of the person controlling the canal, sea outfall or other conduit"). Based on the Constitution of the RSA, Schedule 4 – Part B (Republic of South Africa 1996), stormwater management services in urban areas are the responsibility of municipalities. In reality, municipalities often assign the stormwater management role to road departments (Fishcr-Jeffes 2015).

The RSA's stormwater management legislation includes the Stormwater Management By-law (City of Capetown 2005) and the Management of Urban Stormwater Impact Policy (Roads and Stormwater Department-Republic of South Africa 2009). The latter policy aims to identify measures to "reduce the impact of flooding on community livelihoods and regional economies and safeguard human health, protect natural aquatic environments, and improve and maintain recreational water quality". In this policy, criteria (e.g. pollutant removal targets) for sustainable urban drainage systems are detailed (Roads and Stormwater Department-Republic of South Africa 2009).

The City of Cape Town is an example of a municipality that has developed their own by-laws for improved stormwater management. For example, Clause 3 of the

City of Cape Town's Stormwater Management By-law prohibits discharge of pollutants (anything other than stormwater) into the stormwater system, while Clauses 4 and 5 of the by-law deal with protection of the stormwater system and the prevention of flood risk.

1.5.6 Southeast Asia

While the management of stormwater in Southeast Asia varies across the region, the importance of stormwater management is well recognised as the region faces ongoing rapid growth and urbanisation. Regionally, the Association of Southeast Asian Nations (ASEAN) formed the ASEAN Working Group on Water Resources Management to cooperatively address water management concerns. It has a focus on six key water management issues, three of which relate to stormwater – stormwater management, flood management and water pollution management. The working group publishes progress reports on each water management issue, which are available to summarise progress in each country across the region (Integrated water resources management 2018). Using available records for 2013, the following table (Table 1.1) presents an overview of stormwater management progress for ASEAN member countries based on self-reporting processes to the ASEAN working group.

Singapore is reported to be the most advanced country for sustainable measures, with arguably the first comprehensive low impact development programme in the tropics (Lim and Lu 2016). The Singapore Public Utilities Board provides guidelines for stormwater management, including a code of practice on stormwater drainage (Public Utilities Board 2013) which is supported by additional resources on sustainable approaches, including the "Active Beautiful Clean" waters programme.

Table 1.1 Stormwater management progress for ASEAN member countries

Country	Policy	Legislation	Design manual	National budget allocated to flood management (%)	Flood-prone areas covered by flood warning systems (%)	Flood-prone areas covered by flood monitoring system (%)
Brunei[a]	–	–	–	–	–	
Cambodia[a]	–	–	–	–	–	
Indonesia	Yes	Yes	Yes	0.32	17	30
Laos[a]	–	–	–	–	–	
Myanmar	No	No	No	–	–	–
Malaysia	Yes	Yes	Yes	0.64	10	50
Philippines[a]	–	–	–	–	–	
Singapore	Yes	Yes	Yes	–	100	100
Thailand	No	No	No	1	66	28
Vietnam	Yes	Yes	No	10	65	35

[a] Information was not available or found in reports accessed

However, other countries have also developed detailed guidelines interwoven with sustainable measures. In Malaysia, the Department of Irrigation and Drainage has published the Urban Stormwater Management Manual for Malaysia (Department of Irrigation and Drainage 2018).

1.5.7 UK

Governance of stormwater in the UK is covered in more detail in Chap. 2. Broadly speaking, stormwater management in the UK has been influenced by the 2010 Flood and Water Management Act of England and Wales (2010). This was precipitated by the occurrence of extensive floods in the UK in 2007, which led to the Pitt Review (Pitt 2008) and subsequent changes to building regulations and Planning Policy Statement 25. Schedule 3 of the Flood and Water Management Act details the responsibilities of local authorities for surface water management, which include coordination and liaison with other authorities (e.g. the UK Environment Agency and Internal Drainage Boards) to develop water management plans and adopt a wide range of SuDS. Under this schedule is the required National Standards for SuDS, which include standards for stormwater treatment. From October 2015, all developments of more than one house required approval from SuDS Approval Bodies and existing planning authorities were given responsibility to consider approval of SuDS proposals for ten or more dwellings or equivalent non-residential or mixed development. The respective legislation of Scotland and Northern Ireland are the Flood Risk Management Act 2009 and the Water Environment (Floods Directive) Regulations (Northern Ireland) 2009 (Stormwater Management Limited 2018).

1.6 Future Needs for Improving Liveability

The aims of more recent stormwater management initiatives such as Sponge Cities, LID and WSUD are in response to increased risks and challenges to sustainable water management and, more recently, liveability in urban environments. These are from population growth, urbanisation, changing climate and impacts on sustainable water resources. Challenges to sustaining and improving liveability with respect to stormwater impacts are wide ranging and include known and unknown risks and, where unknown, elucidating or predicting these risks reliably. These include:

- Human hazards from stormwater contaminants from diffuse sources, which may be transported to surface or groundwater resources
- The management and fate of emerging pollutant contaminants such as PFAS that threaten the reuse of stormwater or add to the costs of treatment for their removal

- Lack of reliable prediction of extreme climate events and climate cycling (e.g. the Walker Cycle) that in some countries and states significantly impacts on the approaches needed for water resource management
- The high financial costs associated with implementing retrospectively and, in greenfield developments, state-of-the-art stormwater management for discharge and reuse purposes

In developing and/or socially, politically and economically unstable countries, such as those with high urban population growth and low-level planning capacity, stakeholders face ongoing challenges in attaining or organising the specific resources (human and capital) needed for ongoing, well-organised, stormwater management. A key challenge lies in addressing these matters in the context of raising social expectations, political instability and economic capacity for meeting current and future stormwater management needs while minimising risk to people and infrastructure from flooding.

Even in developed industrialised nations, such as those in North America, Europe and Asia, there is a need to modernise existing systems. Managers need to balance the allocation of available funds for stormwater infrastructure and management with other community requirements (Porse 2013).

References

ACT (2009) Design standards for urban infrastructure. Transport Canberra and city services, ACT Government, Australia. https://www.tccs.act.gov.au/Development_and_Project_Support/pre-development-applications/estate-development-plans/design_standards_for_urban_infrastructure. Accessed 31 July 2018

Adyel TM, Oldham CE, Hipsey MR (2016) Stormwater nutrient attenuation in a constructed wetland with alternating surface and subsurface flow pathways: event to annual dynamics. Water Res 107(Supplement C):66–82. https://doi.org/10.1016/j.watres.2016.10.005

African Progress Panel (2015) Power people planet: seizing Africa's energy and climate opportunities: Africa progress report 2015. ISBN 978-2-9700821-6-3

Agriculture and Resource Management Council of Australia and New Zealand (1996) Draft guidelines for urban stormwater management. Australian and New Zealand Environment and Conservation Council, ACT, Canberra

Ahn JH, Grant SB, Surbeck CQ, DiGiacomo PM, Nezlin NP, Jiang S (2005) Coastal water quality impact of stormwater runoff from an urban watershed in Southern California. Environ Sci Technol 39(16):5940–5953. https://doi.org/10.1021/es0501464

Argue J, Pezzaniti D (2005) Chapter 11-infiltration systems. In: Australian Runoff Quality. Institution of Engineers, Sydney

Aryal R, Vigneswaran S, Kandasamy J, Naidu R (2010) Urban stormwater quality and treatment. Korean J Chem Eng 27(5):1343–1359

Ashley R, Jones J, Ramella S, Schofield D, Munden R, Zabatis K, Rafelt A, Stephenson A, Pallett I (2007) Delivering more effective stormwater management in the UK and Europe: lessons from the clean water act in America. NOVATECH 2007

Austrade (2016) China's sponge city program. https://www.austrade.gov.au/ArticleDocuments/6585/China%20Sponge%20City%20Program.pdf.aspx

Australian Bureau of Statistics (2013) 4602.0.55.003 – Environmental issues: water use and conservation, Rainwater tanks. http://www.abs.gov.au/ausstats/abs@.nsf/Lookup/4602.0.55.003main+features4Mar%202013. Accessed 20 Nov 2017

Australian Drinking Water Guidelines (2011) Australian drinking water guidelines 6 national water quality management strategy. vol 3.4, Updated October 2017. Australian Government, National Health and Medical Research Council (NRMMC), Australia

Australian Government Department of Defence (AGDD) (2016) RAAF base Edinburgh fact sheet. http://www.defence.gov.au/Environment/PFAS/Docs/Edinburgh/FactSheets/EdinburghFactSheetOctober2016.pdf

Ball J, Babister M, Nathan R, Weeks W, Weinmann E, Retallick M, Testoni I (2016) Australian rainfall and runoff: a guide to flood estimation. Commonwealth of Australia, Canberra

Barbosa AE, Fernandes JN, David LM (2012) Key issues for sustainable urban stormwater management. Water Res 46(20):6787–6798. https://doi.org/10.1016/j.watres.2012.05.029

Beijing Capital Group Company Ltd Beijing Capital's First Sponge City PPP Project Signs Contract and Breaks Ground. https://www.bjcapital.com/en/content/details_117_2605.html. Accessed 19 Nov 2017

Birch GF, Drage DS, Thompson K, Eaglesham G, Mueller JF (2015) Emerging contaminants (pharmaceuticals, personal care products, a food additive and pesticides) in waters of Sydney estuary, Australia. Mar Pollut Bull 97(1):56–66. https://doi.org/10.1016/j.marpolbul.2015.06.038

Bradsher J, Wojtala G, Kaml C, Weiss C, Read D (2015) Regulatory foundations for the food protection professional. Springer, New York

Bratieres K, Fletcher T, Deletic A, Zinger Y (2008) Nutrient and sediment removal by stormwater biofilters: a large-scale design optimisation study. Water Res 42(14):3930–3940

Brown R, Ashley R, Farrelly M (2011) Political and professional agency entrapment: an agenda for urban water research. Water Resour Manag 25(15):4037–4050. https://doi.org/10.1007/s11269-011-9886-y

Bureau of Meteorology (2015) Recent rainfall, drought and southern Australia's long-term rainfall decline. Bureau of meteorology, Australian Government. http://www.bom.gov.au/climate/updates/articles/a010-southern-rainfall-decline.shtml. Accessed 31 Dec 2017

Burns MJ, Fletcher TD, Walsh CJ, Ladson AR, Hatt BE (2012) Hydrologic shortcomings of conventional urban stormwater management and opportunities for reform. Landsc Urban Plan 105(3):230–240. https://doi.org/10.1016/j.landurbplan.2011.12.012

Chandana D, Giacomoni MH, Prakash Khedun C, Hillary H, Andrea R, William S, Zechman EM (2010) Simulation of combined best management practices and low impact development for sustainable stormwater management. JAWRA J Am Water Resour Assoc 46(5):907–918. https://doi.org/10.1111/j.1752-1688.2010.00462.x

Chattopadhyay S, Harilal K (2017) Emerging challenges of urban water governance case of Thiruvananthapuram city, India. Research Unit on Local Self Governments, Centre for Development Studies, Thiruvananthapuram

Che W, Zhang W, Liu DK, Gan YP, Lv FF (2011) Characterization of urban runoff from combined sewer and storm sewer catchments in Beijing, China. In: ICPTT 2011. doi:https://doi.org/10.1061/41202(423)251

City of Capetown (2005) Stormwater management By-law (Provincial Gazette no. 6300). https://indigo.openbylaws.org.za/api/za-cpt/act/by-law/2005/stormwater-management/eng.pdf. Accessed 20 June 2018

City of Salisbury (2016) Audit committee meeting: report 3.1.3. http://www.salisbury.sa.gov.au/files/sharedassets/public/website_businesspapers/audit/15112016_-_agenda_and_reports_-_audit_committee_of_council.pdf?BestBetMatch=pfas|9d82872e-4b5e-4c6f-a134-d113dc2dea71|0e93b394-4add-4c5f-8840-a2280124c87f|. Accessed 12 Nov 2017

Cook S, Myers B, Newland P, Pezzaniti D, Kemp D (2015) Pathways for implementation of water sensitive urban design policy in South Australia. Goyder Institute for Water Research, Adelaide

Coombes P, Mitchell G, Breen P (2006) Roofwater, storm water, and wastewater reuse. Austr Run-off Qual:5.1–5.22

Cox TE, Foster MS (2013) The effects of storm-drains with periodic flows on intertidal algal assemblages in ‘Ewa Beach (O‘ahu), Hawai‘i. Mar Pollut Bull 70(1):162–170. https://doi.org/10.1016/j.marpolbul.2013.02.029
Davis AP, Shokouhian M, Sharma H, Minami C, Winogradoff D (2003) Water quality improvement through bioretention: Lead, copper, and zinc removal. Water Environ Res 75(1):73–82
de la Trincheria J, Yemaneh A (2016) Monitoring and management of flowing rain water in Baltic Sea catchment areas, final report. European Commission 7th framework Work programme
de Vriend HJ, van Koningsveld M (2012) Building with nature: thinking, acting and interacting differently. EcoShape, Building with Nature, Dordrecht
Department of Environment and Conservation (2006) Managing urban stormwater: harvesting and reuse. ISBN 1 74137 875 3. New South Wales
Department of Environment and Energy (2002) Introduction to urban stormwater management in Australia Department of Environment and Energy, Australian Government. http://www.environment.gov.au/resource/introduction-urban-stormwater-management-australia. Accessed 8 Jan 2018
Department of Environment and Heritage (2002) Introduction to urban stormwater management in Australia. Environment Australia, Canberra
Department of Irrigation and Drainage Malaysia Percetakan Nasional Malaysia Berhad Urban storm water management manual for Malaysia. http://msmam.com/msma-chapters/. Accessed 22 June 2018
Department of Planning Transport and Infrastructure (2013) Recycled water guideline. Sustainable use of water for infrastructure construction, maintenance and irrigation. vol K-Net Doc: 2531416 Department of Planning, Transport and Infrastructure, South Australia Government
Dhakal KP, Chevalier LR (2016) Urban stormwater governance: the need for a paradigm shift. Environ Manag 57(5):1112–1124. https://doi.org/10.1007/s00267-016-0667-5
Dierkes C, Lohmann M, Becker M, Raasch U (2005) Pollution retention of different permeable pavements with reservoir structure at high hydraulic loads. In: Proceedings of the 10th international conference on urban drainage, pp 21–26, Copenhagen, Denmark
Dillon P, Page D, Dandy G, Leonard R, Tjandraatmadja G, Vanderzalm J, Rouse K, Barry K, Gonzalez D, Myers B (2014) Managed aquifer recharge and Stormwater use options: summary of research findings. Goyder Institute for water research, Technical report 14:13
Dwr Cymur Welsh Water – Combined sewers and combined sewer overflows. https://dwrcymru.com/en/Devmap172.aspx. Accessed 8 Nov 2017
Environment and Communications References Committee (2015) Environment and communications references committee – stormwater management in Australia. Commonwealth of Australia, Parliament House, Canberra ACT 2600
Environment Protection Authority South Australia (2007) EPA guidelines for stormwater management in Mount Gambier. Environment Protection Authority South Australia, Adelaide
Fabricius KE (2005) Effects of terrestrial runoff on the ecology of corals and coral reefs: review and synthesis. Mar Pollut Bull 50(2):125–146. https://doi.org/10.1016/j.marpolbul.2004.11.028
Ferguson BC, Brown RR, Frantzeskaki N, de Haan FJ, Deletic A (2013) The enabling institutional context for integrated water management: lessons from Melbourne. Water Res 47(20):7300–7314. https://doi.org/10.1016/j.watres.2013.09.045
Fisher-Jeffes LN (2015) The viability of rainwater and stormwater harvesting in the residential areas of the Liesbeek River catchment, Cape Town. University of Cape Town, Cape Town
Fletcher TD, Shuster W, Hunt WF, Ashley R, Butler D, Arthur S, Trowsdale S, Barraud S, Semadeni-Davies A, Bertrand-Krajewski J-L, Mikkelsen PS, Rivard G, Uhl M, Dagenais D, Viklander M (2015) SUDS, LID, BMPs, WSUD and more – the evolution and application of terminology surrounding urban drainage. Urban Water J 12(7):525–542. https://doi.org/10.1080/1573062X.2014.916314
Flood and Water Management Act (2010) Flood and Water Management Act. Public General Acts – öElizabeth II chapter 29, The Stationery Office, London

Fox DR, Batley GE, Blackburn D, Bone Y, Bryars S, Cheshire A, Collings G, Ellis D, Fairweather P, Fallowfield H, Harris G, Henderson B, Kämpf J, Nayar S, Pattiaratchi C, Petrusevics P, Townsend M, Westphalen G, Wilkinson J (2007) The adelaide coastal waters study, vol 1. CSIRO, Adelaide

Furlong C, De Silva S, Gan K, Guthrie L, Considine R (2017) Risk management, financial evaluation and funding for wastewater and stormwater reuse projects. J Env Manag 191(Supplement C):83–95. https://doi.org/10.1016/j.jenvman.2017.01.007

Garner E, Benitez R, von Wagoner E, Sawyer R, Schaberg E, Hession WC, Krometis L-AH, Badgley BD, Pruden A (2017) Stormwater loadings of antibiotic resistance genes in an urban stream. Water Res 123:144–152. https://doi.org/10.1016/j.watres.2017.06.046

Gerrity D, Pecson B, Trussell RS, Trussell RR (2013) Potable reuse treatment trains throughout the world. J Water Supply Res Technol AQUA 62(6):321–338

Government of South Australia (2010) Water sensitive urban design technical manual for the greater Adelaide region. Department of Planning and Local Government, Adelaide

Government of South Australia – planning professionals and developers: water-sensitive urban design. https://www.sa.gov.au/topics/planning-and-property/land-and-property-development/planning-professionals/water-sensitive-urban-design. Accessed 20 Nov 2017

Hains S (2009) Stormwater: towards water sensitive cities. The emerging crisis of water in southern Australia. Paper presented at the LGMA National Congress, Darwin, Australia, 26/05/2009

Hamel P, Fletcher TD (2013) The impact of stormwater source-control strategies on the (low) flow regime of urban catchments. Water Sci Technol 69(4):739–745. https://doi.org/10.2166/wst.2013.772

Hatt BE, Deletic A, Fletcher TD (2006) Integrated treatment and recycling of stormwater: a review of Australian practice. J Environ Manag 79(1):102–113. https://doi.org/10.1016/j.jenvman.2005.06.003

Hatt B, Deletic A, Fletcher T (2007) Stormwater reuse: designing biofiltration systems for reliable treatment. Water Sci Technol 55(4):201–209

IANAS (2015) The inter-American network of academies of sciences, urban water challenges in the Americas, A perspective from the Academies of Sciences. IANAS and UNESCO, Tlalpan, Mexico

Institute of Public Works Engineering Australia (2013) Queensland urban drainage manual – third edition. Brisbane City Council, QLD, Australia

Integrated water resources management water management issues. http://aseaniwrm.water.gov.my/. Accessed 26 May 2018

Jayasuriya N, Kadurupokune N (2010) Comparative performance of permeable and porous pavements, in International Conference on Sustainable Built Environment (ICSBE 2010) December 13–14, Kandy, Sri Lanka

Kalmykova Y, Björklund K, Strömvall A-M, Blom L (2013) Partitioning of polycyclic aromatic hydrocarbons, alkylphenols, bisphenol A and phthalates in landfill leachates and stormwater. Water Res 47(3):1317–1328. https://doi.org/10.1016/j.watres.2012.11.054

Kinsella CM, Crowe TP (2015) Variation in rocky shore assemblages and abundances of key taxa along gradients of stormwater input. Mar Environ Res 105:20–29. https://doi.org/10.1016/j.marenvres.2015.01.003

Kjaer AM (2004) Governance. Polity Press. Malden, MA, USA

Legret M, Colandini V, Le Marc C (1996) Effects of a porous pavement with reservoir structure on the quality of runoff water and soil. Sci Total Environ 189:335–340

Li T, Zhang W, Feng C, Shen J (2014) Performance assessment of separate and combined sewer systems in metropolitan areas in southern China. Water Sci Technol 69(2):422–429. https://doi.org/10.2166/wst.2013.732

Li H, Ding L, Ren M, Li C, Wang H (2017) Sponge city construction in China: a survey of the challenges and opportunities. Water 9(9):594

Lim HS, Lu XX (2016) Sustainable urban stormwater management in the tropics: an evaluation of Singapore's ABC waters program. J Hydrol 538:842–862. https://doi.org/10.1016/j.jhydrol.2016.04.063
Mullaney J, Lucke T (2014) Practical review of pervious pavement designs. Clean (Weinh) 42(2):111–124. https://doi.org/10.1002/clen.201300118
Myers B, Beecham S, Leeuwen JAV (2011) Water quality with storage in permeable pavement basecourse. Proc Inst Civ Eng – Water Manag 164(7):361–372. https://doi.org/10.1680/wama.2011.164.7.361
National Research Council (2008) Urban stormwater management in the United States. National Academies Press, Washington, DC
NetRegs-Sustainable urban drainage systems. Environmental guidance, Northern Ireland Environment Agency and Scottish environment protection agency. http://www.netregs.org.uk/environmental-topics/water/sustainable-drainage-systems-suds/. Accessed 19 Nov 2017
Nnadi EO, Newman AP, Coupe SJ, Mbanaso FU (2015) Stormwater harvesting for irrigation purposes: an investigation of chemical quality of water recycled in pervious pavement system. J Environ Manag 147:246–256
Nolde E (2007) Possibilities of rainwater utilisation in densely populated areas including precipitation runoffs from traffic surfaces. Desalination 215(1):1–11. https://doi.org/10.1016/j.desal.2006.10.033
Olsson L, Head BW (2015) Urban water governance in times of multiple stressors: an editorial. Ecol Soc 20(1)
Ontario (1990) Ontario Water Resources Act. R.S.O. Accessed 17 July 2018
Oxford Dictionaries. https://en.oxforddictionaries.com/definition/storm_water. Accessed 10 Oct 2017
Page D, Miotliński K, Gonzalez D, Barry K, Dillon P, Gallen C (2014) Environmental monitoring of selected pesticides and organic chemicals in urban stormwater recycling systems using passive sampling techniques. J Contam Hydrol 158:65–77. https://doi.org/10.1016/j.jconhyd.2014.01.004
Page DW, Peeters L, Vanderzalm J, Barry K, Gonzalez D (2017) Effect of aquifer storage and recovery (ASR) on recovered stormwater quality variability. Water Res 117(Supplement C):1–8. https://doi.org/10.1016/j.watres.2017.03.049
Park M-H, Reckhow D, Lavine M, Stanford B, Rosenfeldt E, Lipsky D, Mofidi A (2017) EDC/PPCP benchmarking and monitoring strategies for drinking water utilities. Vol project #4260. Water Research Foundation, 6666 West Quincy avenue, Denver, CO 80235
Pierre J, Peters BG (2000) Governance, politics, and the state. St. Martin's Press, New York
Pitt M (2008) The Pitt review – learning lessons from the 2007 floods. Cabinet Office, London
Porse EC (2013) Stormwater governance and future cities. Water 5(1):29–52
Public Utilities Board (2013) Code of practice on surface water drainage, 6th edn. (Dec-2011) with amendments under Addendum No.1-June 2013. Drainage Department, Ministry of the Environment, Singapore
Pyne RDG (1995) Groundwater recharge and wells: a guide to aquifer storage recovery. CRC Press, Boca Raton
Republic of South Africa (1996) Constitution of the Republic of South Africa Act No. 108 of 1996. Government of Republic of South Africa, Pretoria
Republic of South Africa (1998) National Water Act, Act no 36 of 1998. Government of Republic of South Africa, Pretoria
Reuters (23/03/2017) Holm-Nielsen, N. (Disaster risk management, World Bank); Nalesso, M. (Inter-American Development Bank (IDB). https://www.reuters.com/article/us-peru-floods-weather-climatechange/perus-deadly-floods-ring-alarm-bell-for-latin-america-cities-idUSKBN16U11B. Accessed 17 June 2018
Roads & Stormwater Department-Republic of South Africa (2009) Management of urban stormwater impact policy (C 58 05/09). City of Cape Town, South Africa

Roy-Poirier A, Champagne P, Filion Y (2010) Review of bioretention system research and design: past, present, and future. J Env Eng 136(9):878–889. https://doi.org/10.1061/(ASCE)EE.1943-7870.0000227

Ryu J, Oh J, Snyder SA, Yoon Y (2014) Determination of micropollutants in combined sewer overflows and their removal in a wastewater treatment plant (Seoul, South Korea). Environ Monit Assess 186(5):3239–3251. https://doi.org/10.1007/s10661-013-3613-5

Schiff K, Bay S (2003) Impacts of stormwater discharges on the nearshore benthic environment of Santa Monica Bay. Mar Environ Res 56(1):225–243. https://doi.org/10.1016/S0141-1136(02)00332-X

Schwammberger P, Walker C, Lucke T (2017) Using floating wetland treatment systems to reduce stormwater pollution from urban developments. Int J GEOMATE 12(31):45–50

Smith ME, Fairweather PG (2016) Effects of proximity to stormwater on the sandy-beach macrofaunal assemblages of metropolitan Adelaide, South Australia. Mar Environ Res 122:76–84. https://doi.org/10.1016/j.marenvres.2016.09.009

Southeast Michigan Council of Governments (2008) Investment in reducing combined sewer overflows pays dividends. vol 48226–3602. Southeast Michigan council of governments, Information center 535 Griswold Street, Suite 300 Detroit, MI

Southern Water, UK - Combined sewer overflows. https://www.southernwater.co.uk/combined-sewer-overflows. Accessed 16 Nov 2017

Stormwater Management Limited (2018) Stormwater management: sustainable urban drainage systems. http://www.storm-water.co.uk/legislation. Accessed 6 June 2018

The World Bank (2017) Urban population (% of Total). https://data.worldbank.org/indicator/SP.URB.TOTL.IN.ZS. Accessed 22 Nov 2017

Tjandraatmadja G, Cook S, Chacko P, Myers B, Sharma A, Pezzaniti D (2014) Water sensitive Urban Design impediments and potential: contributions to the SA urban water blueprint – post-implementation assessment and impediments to WSUD. Goyder Institute for Water Research, Adelaide

Trowsdale SA, Simcock R (2011) Urban stormwater treatment using bioretention. J Hydrol 397(3):167–174. https://doi.org/10.1016/j.jhydrol.2010.11.023

United Nations (2014) World's population increasingly urban with more than half living in urban areas. http://www.un.org/en/development/desa/news/population/world-urbanization-prospects-2014.html. Accessed 22 Nov 2017

United States Environmental Protection Agency (2012a) Terminology of low impact development office of wetlands, oceans, and watersheds United States environmental protection agency, 1200 Pennsylvania Avenue, NW, Washington, DC

United States Environmental Protection Agency (2012b) Benefits of low impact development. Office of wetlands, oceans, and watersheds. united states environmental protection agency, 1200 Pennsylvania Avenue, NW, Washington, DC 20460

Victorian Builders Authority (2014) About stormwater. http://www.google.com.au/url?sa=t&rct=j&q=&esrc=s&source=web&cd=1&ved=0ahUKEwiwif2g7LfXAhVMlpQKHZm0At8QFggpMAA&url=http%3A%2F%2Fwww.vba.vic.gov.au%2F__data%2Fassets%2Fpdf_file%2F0016%2F21265%2FStormwater.pdf&usg=AOvVaw3yqtV_rztobSGNqxqX_Mbw. Accessed 12 Nov 2017

Victorian Stormwater Committee (1999) Urban stormwater: best-practice environmental management guidelines. CSIRO publishing, Melbourne

Walker C, Tondera K, Lucke T (2017) Stormwater treatment evaluation of a constructed floating wetland after two years operation in an urban catchment. Sustainability 9(10):1687

Walsh CJ, Fletcher TD, Burns MJ (2012) Urban stormwater runoff: a new class of environmental flow problem. PLoS One 7(9):e45814

Western Australia Water Corporation (2005) Reuse scheme benefits local shires and the environment. Western Australia water corporation. https://www.watercorporation.com.au/about-us/news/media-statements/media-release/reuse-scheme-benefits-local-shires-and-the-environment. Accessed 1 Jan 2018

Wong THF (2006) Australian Runoff Quality (ARQ). Engineers Australia, Sydney
Wu H, Huang G, Meng Q, Zhang M, Li L (2016) Deep tunnel for regulating combined sewer overflow pollution and flood disaster: a case study in Guangzhou City, China. Water 8(8):329
Yang J (2016) Integrating sustainable storm water management and urban planning -a comparative case study between Sponge city in China and climate proof city in the Netherlands. Msc thesis LUP80436, Wageningen University, Netherlands
Zhang S, Pang S, Wang P, Wang C, Han N, Liu B, Han B, Li Y, Anim-Larbi K (2016) Antibiotic concentration and antibiotic-resistant bacteria in two shallow urban lakes after stormwater event. Environ Sci Pollut Res 23(10):9984–9992. https://doi.org/10.1007/s11356-016-6237-9

Chapter 2
Stormwater Harvesting and Flood Mitigation: A UK Perspective

Doug Warner, Kathy Lewis, and John Tzilivakis

Abstract The UK is defined by the World Resources Institute as subject to 'medium to high' levels of water stress and as water scarce per capita. Annual rainfall is spatially variable, being in excess of 1000 mm per annum in parts of the north and west of the country, to below half that in areas of the south and east. Potential threats to water supplies in the UK are severe in the south-east of the country, including London. Most areas of the UK have experienced flooding in response to increased storm frequency and ferocity. A criticism of water management in the UK is an over-reliance on a centralised supply system, coupled with a failure to expand the uptake of small-scale locally implementable techniques such as rainwater harvesting and grey water recycling. This represents both a risk to meeting supply demand and a potential hindrance to the mitigation of flood risk within urban environments. The UK has the opportunity to learn from the Australian system of ecologically sustainable design that includes strategies to supplement or substitute supply from a centralised system. It is discussed in this and subsequent chapters.

Keywords Centralised water supply system · Climate change · Drought · Fluvial flooding · Green roof · Intra-urban flooding · Rainwater harvesting · Surface run-off · Water stress indicator

2.1 Overview

The United Kingdom (UK), despite prolific rainfall in the north and west of the country, is defined by the World Resources Institute (2013) as subject to 'medium to high' levels of water stress and as water scarce per capita, at 1400 m^3 $year^{-1}$ per person in England and Wales. This is rated as 'low availability' by Griggs et al. (1997). The UK and Australia are currently categorised within the same water stress

D. Warner (✉) · K. Lewis · J. Tzilivakis
Agriculture and Environment Research Unit (AERU), University of Hertfordshire, Hatfield, Hertfordshire, UK
e-mail: d.j.warner@herts.ac.uk

V. Jegatheesan et al. (eds.), *Urban Stormwater and Flood Management*, Applied Environmental Science and Engineering for a Sustainable Future,
https://doi.org/10.1007/978-3-030-11818-1_2

level grouping (0.2–0.5) defined by Pfister et al. (2009). Table 1.1 demonstrates the differential as a negligible 0.007. Gassert et al. (2013) classify water stress within the UK as being 'moderate'. Dickie (2006) highlights potential threats to water supplies in the UK as being severe in the south-east, and moderate in most other regions, citing London as the *largest significantly groundwater dependant city in the developed world.* The UK experiences sizeable variation in precipitation over relatively small spatial scales (Fig. 2.1a).

Rainfall is typically below 600 mm year^{-1} in East Anglia in contrast to being in excess of 2000 mm year^{-1} in the north-west (Met Office UK 2016). Broadly speaking, the regions with the lowest rainfall such as the south and east, including London, are the most densely populated parts of the country and have the greatest demand on resources (Parsons et al. 2010; Wheater and Evans 2009). The disparity between consumption and availability has produced what are termed 'water-rich' and 'water-poor' regions within the UK (Defra 2008). The UK continues to increase in population, with land per capita in 2016 equal to 4 km^2 per 1000 persons in the UK, compared to 363 km^2 in Australia (Table 2.1).

Further pressure on the water supply is exerted by abstraction for agricultural activities, a greater frequency of seasonal low river flow associated with drought and increased urbanisation. Battarbee et al. (2012) consider these factors to be the greatest challenge for water management in the UK. The expansion in population, coupled with the associated increase in urban areas and consequential coverage by water-impermeable surfaces, exposes the system to a further risk, urban flooding. The increase in severe weather events in the UK, like short but intense periods of

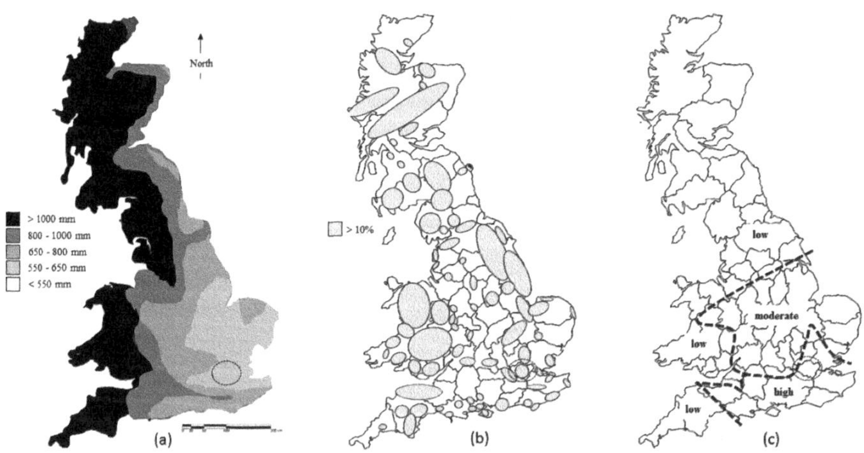

Fig. 2.1 (**a**) mean annual UK rainfall (mm year^{-1}) (adapted from UK Meteorological Office 2018); (**b**) indicative areas (shaded) where greater than 10% of residential dwellings are at risk to flooding, (adapted from Environment Agency 2018); (**c**) proportion of drinking water supplied by groundwater (adapted from British Geological Survey 2015). Dashed circle denotes London

Table 2.1 Summary of key water stress indicators for the UK and Australia (Growing Blue 2016; Pfister et al. 2009; World Bank 2018). Figures in brackets refer to global ranking

	UK	Australia
Water stress level	0.395	0.402
Growth in population (% $year^{-1}$)	0.6	1.6
Potential renewable freshwater per person (m^3 $person^{-1}$ $year^{-1}$)	2864.0	18372.0
Percent water dependency	1.4	N/A
Groundwater abstraction (as percent of potential water recharge)	25.5	3.1
Water extracted – municipal ($m^3 \times 10^9$ day^{-1})	2.1 (34)	3.5 (28)
Water extracted – industrial sectors ($m^3 \times 10^9$ day^{-1})	7.2 (18)	2.4 (31)
Water extracted – agricultural sector ($m^3 \times 10^9$ day^{-1})	0.3 (112)	18.0 (28)
Water footprint per person (m^3 $person^{-1}$ $year^{-1}$)	1245.0 (74)	139253.0 (52)
Persons per km^2	273.0 (172)	3.0 (4)
Arable land (ha $person^{-1}$)	0.1 (128)	1.9 (1)

precipitation, has escalated the frequency of flooding in towns and cities (Fig. 2.1b), where ageing sewerage and drainage systems do not have the capacity to adapt to such events (Curtis and Cooper 2009; Fewkes 2012; Wheater and Evans 2009). The Foresight Future Flooding project (Evans et al., 2004a, b) concluded that the UK faces a significant increase in flood risk by the year 2080, although events since the report was published suggest this risk has already been realised, more than 60 years early. Neither is the concept of flood risk in the UK new. As Wheater and Evans (2009) point out, the effects and future risks have been documented for over 40 years. Despite this, over four decades later the Pitt Review (2008) examination of the 2007 UK floods highlighted a lack of overall responsibility for urban water management in the UK and the need for it to be addressed. There is evidently an urgent need to implement water management strategies within the UK that both mitigate flood risk but also enhance supply system resilience to increased drought frequency.

2.2 Impacts of Stormwater and Flooding

Floods within urban environments have been well documented in the UK in recent years, with economic costs running into millions of pounds sterling. Instances in the cities of Hull, Manchester and Sheffield and the town of Carlisle in Cumbria are examples. A climate and socio-economic investment scenario-based approach taken by Evans et al. (2004a, b) highlighted climate change, fluvial flooding due to catchment land use and 'intra-urban' flooding as key drivers for urban flood risk within the UK. Intra-urban floods originate from within the urban environment itself.

Catchment and coastal-based flooding describes where stormwater flows originate from outside the urban area, but the consequences are realised within it. This mechanism includes river-derived or fluvial floods.

2.2.1 *Fluvial Flooding*

Where a watercourse is in close proximity to or proceeds through an urban environment, fluvial flooding is a risk. The fluvial mechanism depends on the catchment and the soil type and geology within that catchment (infiltration and water holding capacity), plus topographical features such as gradients and the potential for volumes of stormwater to be channelled rapidly into watercourses from, e.g. upland areas (Bowker et al. 2007; Wheater and Evans 2009). Such events result in flash flooding that lasts periods of days or weeks before subsidence (Bowker et al. 2007). These events are more difficult to predict (Curtis and Cooper 2009) as illustrated by the flooding of Carlisle, north-west England, in 2016. Another less documented consequence, but which is applicable to areas where water supplies are more heavily dependent on aquifers, is the reduction in frequency of low river flow and subsequent reduced groundwater recharge (Wheater and Evans 2009). The predictability level, speed of flood onset and duration are all key variables in determining the impact of a flood (Pitt Review 2008). Many catchments in the UK include areas naturally susceptible to flooding after prolonged winter rainfall. Soils reach field capacity and run-off into water courses is prolific and rapid. The variability in land topography, land management and annual rainfall within the UK and therefore flood risk is however considerable even over relatively small spatial scales. It increases the complexity of the response of the system to stormwater and how strategies to manage and mitigate the associated flood risk may be optimised (Bowker et al. 2007; Curtis and Cooper 2009; Wheater and Evans 2009).

Fluvial flooding is further influenced by run-off from within the urban environment, in addition to the surrounding land uses within the catchment. In 2016, after severe flooding in Cumbria, north-west England, the UK Government announced greater investment in flood alleviation schemes. Such schemes target containment of floodwater, but not necessarily a reduction in volume. Emphasis on reducing the contribution of the urban environment to fluvial flow, and promoting a combination of strategies to improve water retention within the urban environment through greater use of 'green roofs' and sustainable drainage systems (Sustainable Drainage Systems), is promoted by several authors (e.g. Fewkes 2012; Jones and MacDonald 2007; Maltby 2012; Whatmore et al. 2010; Wheater and Evans 2009). Improvement to urban water retention complements mitigation of the second flood pathway, intra-urban flooding.

2.2.2 *Intra-Urban Flooding*

The 'intra-urban' system is characterised by urban flooding from rainwater derived and/or mechanisms contained solely within the urban environment itself, for example, surface run-off from roofs and pavements (pluvial flooding) or rising groundwater levels (Bowker et al. 2007; Environment Agency 2013a; Evans et al. 2004a, b; Wheater and Evans 2009). The replacement of porous vegetated areas with impermeable, low-absorbency hard surfaces increases overland flow and run-off (Curtis and Cooper 2009), rendering the natural storage capacity within soils ineffective and resulting in the discharge of high volumes of water within a short time period (Bowker et al. 2007; Wheater and Evans 2009). In the UK, run-off is typically transported via a network of 'surface drains' (also termed storm sewers), shallow pipes just below the surface level, directly into watercourses. When excess volumes of rainwater enter surface drains of insufficient capacity to remove the excess water efficiently, a surcharge or backflow of water occurs as the quantity exceeds the capacity of the drainage pipe. The continued build-up of pressure causes the lifting of manhole covers and discharge onto pavements and roads. This combines with the accelerated overland flow due to the hard surfaces, resulting in accumulation of rainwater and subsequent flooding. If fluvial flooding is in progress, the drainage of surface drains is further inhibited. Extreme summer rainfall events also have the potential to exert an impact of this nature, even in areas where annual rainfall and excess winter rainfall are lower. Such floods may be more localised in effect, but their influence within the urban environment is, nonetheless, important. Their rapid and localised effect also makes them difficult to forecast (Curtis and Cooper 2009). The risk of intra-urban flooding is exasperated by the deterioration in efficiency of ageing drainage systems in the UK through blockage or sedimentation (Curtis and Cooper 2009; Fewkes 2012; Wheater and Evans 2009). Increased urbanisation and overland drainage into watercourses further increase the risk of catchment run-off and fluvial flooding. New building developments in the UK are now required not to increase peak flow in watercourses receiving any stormwater flows such that 'the developed rate of runoff into a watercourse should be no greater than the undeveloped rate of runoff for the same event' (Environment Agency 2013b). These requirements mitigate intra-urban flooding within new construction zones. Attention is required for areas developed before 2007, which applies to a significant proportion of urban development. Future predictions of increased frequency in storm events under climate change scenarios, coupled with increased urbanisation, require mitigation strategies to be implemented as a matter of urgency.

2.3 Current Management of Stormwater and Flood Risk Mitigation

Reducing overland flow within urban environments is highlighted as critical to safeguard water supplies and quality. It is also necessary to mitigate flood risk. Policy associated with community planning and building construction to mitigate flood risk is the responsibility of the Department for Communities and Local Government (DCLG). The DCLG (2009) Planning Policy Statement PPS25, now superseded by the New Planning Practice Guidance (HM Government 2014), outlines a five-step flood management risk hierarchy for the UK (assess, avoid, substitute, control, mitigate) with recommendations that it be incorporated into regional and sub-regional spatial strategies, local development documents (LDDs) and sustainability appraisals as a strategic approach to mitigating flood risk. In new developments, all five steps of the process can be implemented. Steps 1 and 2 (assess and avoid) evaluate and target areas of low flood risk for the selection of development sites. Step 3 (substitution) specifies that the development type not be vulnerable to the predicted flood risk for a given site. Areas of high risk, for example, require 'water-compatible development' not impacted by flooding (HM Government 2014). Steps 4 and 5 (control and mitigation) are applicable to all areas, including existing developments where flooding has increased in risk. The larger-scale implementations of SuDS deemed critical by Evans et al. (2004a, b) and Ainger et al. (2009) are a component of Step 4 (control) and are discussed in more detail in Chap. 3. Measures applicable

Table 2.2 Measures to increase supply resilience and mitigate flood risk, proposed for the UK

Measure	Description	Spatial scale
Catchment-scale storage level	Reservoirs	Catchment
Land use planning and management	Developments to be sited in areas of low risk	Catchment/town/individual estates within towns
Engineered flood storage	Diversion of rainwater from developed areas into storage zones then released after flood subsides	Town/individual estates within towns
Urban storage above and below ground	Temporary storage of floodwater via detention basins or retention basins for longer-term storage	Town/individual estates within towns
Main drainage form, maintenance and operation	Pipes with greater diameter (hold greater quantities of rainwater), slows release into water courses, reduces risk of 'back-up' within the system	Town/individual estates within towns
Development of urban areas, operation and form	Replace existing hard surfaces with permeable surfacing	Town/individual estates within towns/individual properties
Urban source control and above ground pathways	SuDS, storage reservoirs, limitation on area of hard surfaces, porous (permeable) pavements, rain gardens, infiltration planters, trees and tree boxes, green roofs, rainwater harvesting	Town/individual estates within towns/individual properties

to the intra-urban system (Battarbee et al. 2012; Evans et al. 2004a, b; Wheater and Evans 2009) operate at various spatial scales, summarised in Table 2.2.

Wheater and Evans (2009) cite reservoir construction as a means of 'detention storage' or the construction of porous 'soakaways' onto which stormwater should be channelled to reduce the volume and rate of surface run-off. Evans et al. (2004a, b) prioritises 'Building development, operation and form' that includes the permeable surfacing of existing expanses of hard surfaces, for example, car parks, and rainwater harvesting (RWH). It was first prioritised under the Foresight Futures (Evans et al. 2004a, b) socio-economic scenarios 'Global Sustainability' scenario (Wheater and Evans 2009). The increased utilisation of vegetation or materials such as gravel or porous paving that function in a similar capacity, or underground box or stone-filled soakaways, reduces surface run-off and crucially increases the period between the extreme rainfall event and the discharge of water into the drainage system. UK legislation now prohibits domestic households from the replacement of grassed garden areas with hard surfaces, without planning permission (DCLG 2008b). The 'urban source control and above ground pathways' mechanism also features prominently (Wheater and Evans 2009). Source control, the reduction of flow, includes the increased utilisation of 'green roofs'. Konig (1999) estimates that green roofs are able to intercept 50% or more of rainfall, although they are only currently installed on a limited number of new builds where the plant *Sedum* (subtribe Sedinae), a stonecrop species, is typically used. At the broader spatial scale, green roofs offer the potential to create habitats (subject to the use of appropriate plant mixtures), to mitigate habitat fragmentation and enhance biodiversity (Gedge et al. 2009). Such strategies have the advantage of adapting existing redundant space, where space may be a limiting factor, in densely populated urban environments. The appropriate planning of SuDS in the UK is cited by O'Brien (2014) as having value to amphibian conservation, including the Red Data Book-listed great crested newt (*Triturus cristatus*), afforded legal protection under European legislation. Another method to reduce flow into surface drains from roof space is through RWH. Rainwater harvesting may deliver benefits on two fronts with respect to key issues facing parts of the UK. Threats to water supply and flood risk may both potentially be mitigated by increased uptake of alternative non-mains supply systems such as RWH (Fewkes 2012; Li et al. 2010; Palla et al. 2017).

The potential for harvesting rainwater to reduce storm run-off and provide harvested rainwater for domestic, non-potable uses in the UK is noted by several authors (e.g. Environment Agency 2010; Fewkes 2012; Melville-Shreeve et al. 2016a, b, c; Wheater and Evans 2009), with benefits to be realised, particularly in the more water-scarce south-east of the UK. Both Brown et al. (2005) and the Environment Agency (EA) (2010) note that Germany and Australia have implemented successful strategies, considering both as models that the UK should aim towards. Germany has the greatest uptake of RWH in Europe largely due to incentives from funding made available by local government (Campasino et al. 2017). According to Schuetze (2013) approximately one third of new buildings in Germany are fitted with RWH systems for use in non-potable applications. Contrary to this situation, as far back as 2004, Berndtsson expressed concern that the UK had fallen behind other countries in the uptake of domestic RWH on a large scale, although Hassell (2005) disputed this observation. A decade later however, despite the concerns raised by Berndtsson (2004), RWH uptake in the UK had not, according to Fewkes (2012), progressed significantly. While modern RWH systems are available in the UK, this availability is considered relatively recent by Campasino et al. (2017). Codes and standards that would facilitate RWH uptake in the UK have also been absent until recently, unlike other parts of Europe (Campasino et al. 2017; De Gouvello et al. 2014; Domènech and Saurí 2011). On a positive note, UK researchers are instrumental in the development and promotion of novel pumping systems in RWH systems as a means to reduce energy consumption (e.g. Ward et al. 2011), one of the potential barriers to uptake, as discussed in more detail in Chap. 6. Further, several European countries have incorporated RWH into systems to control urban run-off (Campasino et al. 2017; Iveroth et al. 2013; Palla et al. 2017), as discussed further in Chap. 4.

Although the case for RWH in the UK and in the south-east in particular appears strong, there is at present limited uptake. Systems in the UK mostly supplement mains supplies in commercial installations for non-potable applications (Fewkes 2000, 2012). This is despite apparent receptivity to their implementation from a water-saving perspective (Ward et al. 2013). What then are the main barriers to more wide-scale adoption in the UK? Ward et al. (2012) believe a key barrier is a lack of stakeholder support, with too great an emphasis on what they term 'top-down' support for the technology. They identify three strategic areas to increase the use of RWH: further product development, capacity building to address gaps in expertise (FAO 2004) and a commitment from relevant institutions to implement water-saving technologies. Parsons et al. (2010) identify five key areas that they consider a hindrance to wide-scale uptake of RWH in the UK: inadequacies within regulation and institutions, excessive cost, limited incentives, a limited available knowledge base and a lack of support within the construction industry. Cost and maintenance and barriers from the consumer perspective are also identified by Fewkes (2012), Liu et al. (2007), Ward et al. (2013) and Wheater and Evans (2009). Excessive cost, namely, the requirement to pump water, has been addressed in part by the development of novel RWH systems that utilise low energy pumping systems (Melville-Shreeve et al. 2016c). The multiple institutions of relevance to the governance of water supply in the UK and lack of overall authority attributed to any one of them (Pericli and Jenkins 2015; Pitt Report 2008) is a further barrier.

2.4 Water Supply and Governance in the UK

As highlighted by Campasino et al. (2017), the cost of potable water is a key driver in determining the demand for the installation of RWH systems in many European countries. This in turn impacts on the potential to mitigate flooding and enhance the liveability of a given urban area. Understanding the method of governance of the water sector in the UK is instrumental in understanding the uptake, or lack of, potential flood mitigation strategies such as RWH systems. The water sector in the UK is managed differently depending on the devolved government. Mains supplied water in England and Wales is the responsibility of around 20 private water companies, also known as water service providers (WSPs), a result of privatisation of the public regional authorities in the late 1980s (Wentworth 2012). The water sector remains publically owned in Scotland and Northern Ireland. Mains water is distributed via a centralised system from a combination of groundwater (abstraction from aquifers) or surface waters (reservoirs and rivers) in varying proportions, depending on the region or water resource zone (Environment Agency 2013a). A recent review of domestic water usage in England and Wales by Pericli and Jenkins (2015) draws similar conclusions to the Pitt Review (2008), listing numerous governmental and non-governmental influences encapsulating multiple organisations and departments. Examples include the Consumer Council for Water (CCW), the Drinking Water Inspectorate (DWI), the Water Services Regulation Authority (Ofwat) and Water Wise, in combination with the Department of Energy and Climate Change (DECC), Department for Environment Food and Rural Affairs (Defra) and the Environment Agency (EA). Critically, each has an influence but none has overall control or responsibility. After privatisation of the public regional water authorities in 1989 in England and Wales, the control of pollution became the remit of the National Rivers Authority (NRA), later absorbed within the Environment Agency. Economic regulatory responsibility is currently undertaken by Ofwat, while the monitoring of water quality and drinking water safety is the responsibility of the DWI. The assessment of compliance with quality and environmental standards is undertaken annually (Defra 2006). In summary, there are multiple organisations with no overall authority operating in a somewhat disjointed manner. This, to a degree, hinders the promotion and implementation of flood mitigation strategies such as RWH, despite the potential water shortages threatening the south-east of the UK. A further variable has been the influence of the European Union via a Directive style form of governance.

Applicable legislation within the UK has been derived mainly from Directives implemented in Europe as a whole, although it is acknowledged that the decision by the UK to leave the European Union will mean it is no longer subject to this form of governance. The European Commission Drinking Water Directive (Council Directive 98/83/EC) of 1998 sought to protect water destined for human consumption from contamination and to protect human health. The most recent amendment to the Directive was in October 2015 (European Commission 2015/1787). In response to Council Directive 98/83/EC, the UK Water Industry Act (1999) makes it an offence to 'pollute or potentially pollute' water sources for human consumption. The Act also legally protects groundwater from abstraction 'in excess of reasonable requirements' or its deliberate wastage from wells and boreholes. A second Europe-wide Directive, the European Water Framework Directive (WFD) (European Union 2000; Defra 2014), implements a broad ecosystem-based approach to water resource management. Preventing over-abstraction and contamination from surface run-off and achieving surface waters of 'good ecological status' are core objectives (Kallis and Butler 2001). In terms of decreasing demand, the UK Water Act (2003) drives the 'sustainable use of water resources' and the 'promotion of water conservation'. The Water Supply Regulations (1999) ensure that harvested rainwater and potable water supplies remain entirely separate prohibiting, for example, joined pipework or the potential for backflow. More recently, the UK Flood and Water Management Act (2010) prohibits the automatic connection of drainage pipes from roofs to storm sewers in new building developments. Fewkes (2012) considers this a positive step forward and a factor likely to promote SuDS.

Most UK environmental legislation implemented in recent years has been derived from European Union directives that are then transposed into UK law (Lang and Schoen 2016), as illustrated by the WFD, which has instigated major improvements in water quality (Kallis and Butler 2001). Whether regulations on, for example, water quality are maintained will be the decision of the UK Government (Lang and Schoen 2016). The European Union 'Circular Economy Package' policy introduced in 2015 committed the European Union to improve environmental performance across multiple sectors. One of the performance measures was water recycling. There is concern that without the European Union, the UK will fall further behind in the implementation of water recycling and increasing the utilisation of RWH and grey water (GW) systems. It may, however, represent a significant opportunity, if countries such as Australia where implementation is at a more advanced stage (Fewkes 2012) are followed.

2.4.1 Water Consumption in the UK

The promotion of water efficiency in new buildings by using strategies such as RWH falls within the remit of the Code for Sustainable Homes (DCLG 2008a, 2010). All homes built post April 2007 in the UK are required to be rated against this code which stipulates minimum environmental performance in six main areas, including water efficiency and surface water management. A grading system of between 1 and 6 is employed, reflecting least and greatest sustainability, respectively. The inclusion of water-saving technologies in a building that reduce internal per capita consumption (PCC) of water (measured in L person^{-1} day^{-1}) increases the sustainability rating of a given dwelling. Within the Code, water consumption categories range from 125 L person^{-1} day^{-1} (Levels 1 and 2) to 105 L person^{-1} day^{-1} (Levels 2 and 3) down to 80 L person^{-1} day^{-1} at the most optimal levels, 5 and 6. The inclusion of RWH in a new build may increase the sustainability score of the building to the maximum of 6 if the PCC is reduced to below 80 L person^{-1} day^{-1}. Part G (sanitation, hot water safety and water efficiency) of the Building Regulations (2016) stipulate maximum consumption to be 125 L person^{-1} day^{-1}, lower than the Defra (2011) target of 130 L person^{-1} day^{-1}. To put this into perspective, current UK domestic water use is estimated at 150 L person^{-1} day^{-1} (Defra 2008; Hunt and Rogers 2014). Compared to the rest of Europe, this places the UK somewhere in the middle, with consumption in Europe as a whole ranging between 100 L person^{-1} day^{-1} (Estonia) and 294 L person^{-1} day^{-1} (Romania) (Defra 2008). The UK's nearest neighbours, Ireland and France, have daily per capita consumption of 190 L $person^{-1}$ day^{-1} and 150 L $person^{-1}$ day^{-1}, respectively. Defra (2011) aims to reduce domestic consumption to 130 L $person^{-1}$ day^{-1}. The main contributors to domestic consumption in the UK include lavatory flushing (30%), washing machines and dishwashers (21%), watering gardens (7%) and bathing (12–21%) (Water Wise 2012). Only 4% is used for drinking. Of these, lavatory flushing, gardening use and possibly clothes washing depending on consumer flexibility could be sourced from recycled GW or harvested rainwater. In commercial buildings, environmental performance incorporating water efficiency is rated using the Building Research Establishment Environmental Assessment Method (BREEAM 2016). However, this is not compulsory.

As highlighted by Campasino et al. (2017) and Godskesen et al. (2013), the cost of potable water is a key driver in determining the demand for the installation of RWH systems in many European countries. Water charges in England and Wales currently take on two forms: a flat rate based on the size of the dwelling as a function of size or direct consumption as measured by a water meter. Metering has been mandatorily introduced into areas designated by the Secretary of State as being 'seriously water stressed' (Wentworth 2012). To date, it has been utilised by Southern Water within the south-east of England, although according to Wentworth (2012) the Institute of Chartered Engineers and Water Wise recommend that all properties within the UK be metered. The Flood and Water Management Act (2010) permits variable tariffs to be implemented. Currently these are limited to 'social tariffs' for consumers who are unable to pay standard rates. Seasonal tariffs are suggested as a means to reduce demand, so that rates increase when supply is under stress. Alternative supplies, for example, RWH or recycled GW, would in theory be an attractive proposition if compulsory metering is expanded and charges increase.

2.4.2 *Current Threats to the UK Water Supply*

2.4.2.1 Depletion of Sources

In 2008, the then UK chief scientific officer John Beddington referred to 'the perfect storm' of global events, in which increased resource demand coupled with decreased resource availability would lead to shortages of water, food and energy by the year 2030 (Beddington 2008). Two key risks to the UK water supply are identified by Charlton and Arnell (2011) who also estimate the percentage contribution of these key drivers to increased future pressure on water resources: (1) a change in demand, responsible for 56%, and (2) a reduction in supply due to climate change, responsible for 37%. Defra (2008, 2011), Environment Agency (2009) and Water UK (2016) identify a need to address potential water shortages within the UK, especially in the south eastern region, and to account for the potential impact of climate change. The HM Government's Sustainable Development Strategy (2005) 'Securing the future' identifies water resource use (total abstractions from nontidal surface and groundwater sources) and domestic water consumption per head as indicators of sustainable resource consumption. In Wales, an area of the UK less prone to drought, sustainable water consumption and water use efficiency are prioritised in the Welsh Assembly Government's Strategic Policy Position Statement on Water (2011). Comparable strategies exist in Scotland (e.g. Scottish Water 2010) and Northern Ireland (Department for Regional Development Northern Ireland 2016). Increases in demand have intensified pressure on the mains water supply throughout Europe, including countries with relatively lower populations and larger land area per capita, for example, Ireland (Li et al. 2010). The population of England is over 10 times that of Ireland but with a similar area of land and living standards. The pressure on the centralised UK water supply network is, therefore, significant. The current shortage of housing stock within the UK, particularly in London and the south-east, and the need for future housing development, will serve only to increase this pressure in the future. Climate change, the second key driver identified by Charlton and Arnell (2011), is also likely to impact London and the south-east the most.

The most water-scarce region of the UK, the south-east of England, is predominantly aquifer fed (Environment Agency 2013a). The vulnerability of the UK to drought, the south-east in particular, is predicted to increase under climate change (Charlton and Arnell 2011; Environment Agency 2013a). Several reports (Defra 2008; 2011; Environment Agency 2009) highlight a significant risk to future water supplies in southern England and the need for measures to be taken by WSPs to ensure they continue to meet demand under changing circumstances. An obligation under the regulatory process in England and Wales is for WSPs to periodically review the investment needed in order to maintain supplies and conform to environmental standards (Wentworth 2012). The review constitutes the production of a Water Resources Management Plan proposing a strategy for a minimum period of 25 years into the future (Environment Agency 2015). A review of recent draft Water Resources Management Plans for the period 2010–2035 by Charlton and Arnell (2011) highlighted the considerable spatial variability in future supply-demand pressure across the UK, due to climatic variables and the highly variable distribution of the population. The increased frequency of drought and extremes in precipitation has reduced the capacity for water infiltration and subsequent aquifer recharge, also resulting in greater quantities of rainfall proceeding through the water cycle as surface run-off (Environment Agency, 2013a).

The Water Resources Strategy for England and Wales (Environment Agency 2009) predicts a decrease in river flows of up to 80% in the south-east by the year 2050. This is coupled with a 35% increase in demand under an 'uncontrolled demand' scenario that assumes 'business as usual', with no further increase in water-saving strategies being implemented. The planning of security of public water supply is implemented at spatial scales defined by water resource zones (Wentworth 2012). The supply versus demand balance of water resources for a given zone is such that all consumers within that zone receive identical levels of supply reliability and are subject to the same potential risk of supply failure. A worse-case scenario under Security and Emergency Measures Direction requires that supplies are maintained at a minimum of 10 L person^{-1} day^{-1}, not sustainable for prolonged periods from the perspective of the consumer (Environment Agency 2015). In calculating the security of supply, WSPs are required to incorporate water availability due to climate change impacts on river flow and groundwater recharge into their 'headroom methodologies' (Charlton and Arnell 2011). This defines the difference between the 'total water available for use' and 'water demand at any given time'. For a given water resource zone, impacts are calculated for three scenarios (dry, medium and wet) projected to the year 2035 for current Water Resources Management Plans. A deficit was predicted for around 50% of zones (Climate Change Committee Adaptation Sub-Committee 2012). Charlton and Arnell (2011) express concern about the quality of the predicted outputs, citing significant variability and uncertainty between scenarios, especially in the more vulnerable resource zones. If indeed this was the case, planning would be informed inaccurately, leading to further risks to supplies. Equally, Charlton and Arnell (2011) criticise the plans further for a lack of adaptation in practice, noting the absence of catchment or aquifer specific models for many water resource zones, and a lack of resources and data with which to run such models. Failure to invest sufficiently in drought mitigation strategies is noted by Water UK (2016). More complex approaches to modelling are currently under development, for example, by Abdellatif et al. (2015) and Diao et al. (2014), but are presently in the theoretical or pilot phases. A recent review of the artificial neural networks (ANN) approach currently under development by authors such as Abdellatif et al. (2015) is given by Kasiviswanathan and Sudheer (2017). They find that the incorporation of uncertainty into the models still requires further development, citing a lack of consideration for the interaction between multiple factors of uncertainty in combination. Such factors are currently considered in isolation to each other.

The UK Environment Agency and Defra's (2008) 'Future Water' strategy promotes a 'twin-track' approach (demand- and supply-based options used simultaneously) to future water management and climate change adaptation. More recently, Water UK (2016) supported this approach as the most appropriate way forward. Contrary to the Government preference, Arnell and Delaney (2006) found many water company Water Resources Management Plans expressed bias towards supply-type measures, such as reservoirs or transferring water between resource zones. Demand-type measures, i.e. benchmarking consumption, RWH and recycling GW, were not prioritised. Fourteen of the draft Water Resources Management Plans reviewed by Charlton and Arnell (2011) gave prominence to either increasing the capacity of existing reservoirs or the creation of new reservoirs entirely. Further, it was noted that emphasis was given to larger centralised reservoirs, rather than promoting small local scale strategies (e.g. RWH) within urban environments. Reducing demand is, according to Pericli and Jenkins (2015), not in the interest of WSPs in England and Wales, whose customers pay for the water they consume. In effect, according to Charlton and Arnell (2011), 56% of supply-demand pressure is currently addressed by bias towards increasing supply alone, while 37% of pressure risks mitigation via strategies based upon potentially inaccurate data modelling. The importance of decreasing pressure on demand within the UK is, however, highlighted by a number of strategic reports (for example Defra 2008; Environment Agency 2009) and reviews (Charlton and Arnell 2011; Fewkes 2012; Pericli and Jenkins 2015). Charlton and Arnell (2011) estimate that a 6% decrease in demand is achievable in the UK by sustainability reduction (reductions required in order to comply with enhanced environmental standards). The Water Resources Strategy for England and Wales (Environment Agency 2009) predicts this demand may decrease by up to 15% under a 'sustainable behaviour' scenario, one that includes utilising strategies to harvest rainwater that also complements flood reduction risk.

2.4.2.2 Contamination

A further threat to aquifer-derived supplies includes groundwater quality deterioration (HM Government 2005; Defra 2011), of particular importance in the south-east of England where groundwater is the main source of potable water within the catchment (Fig. 2.1c). Groundwater in urban environments is cited by Wheater and Evans (2009) as contaminated beyond levels that render it suitable as a potable resource, with London highlighted as being at particular risk. For example, the EU drinking water limit for pesticides is 0.1 micrograms per litre. Emerging contaminants such as the molluscicide metaldehyde, available for domestic application in the UK, have been found to be close to this limit in a number of catchments (Stuart et al. 2011). The Water Industry Act (1999) makes it an offence to 'pollute or potentially pollute' water sources for human consumption. One potential source of contamination arises from diffuse pollution via the flow of rainwater across hard surfaces within urban environments. Oil, heavy metals and chemical residues accumulate, and where this then moves onto permeable soils, there is subsequent infiltration into groundwater (Wheater and Evans 2009). In the UK, the treatment of paved areas with the herbicide glyphosate is standard practice, although this typically breaks down within 3 or 4 days of application (Defra 2015). Along with glyphosate, the secondary product AMPA is another potential issue due to accumulation in groundwater. The restriction of surface flow across paved areas and through urban environments is, therefore, also important in reducing the risk of contamination to groundwater supplies.

Climate change is predicted to increase the risk of pollution to water supplies. Increased drought frequency and low river flows have the potential to escalate algal blooms in fluvial systems, while the dilution of pollutants such as nitrates and phosphates is diminished (Whitehead et al. 2013). According to Whitehead et al. (2009), flash flooding will increase the frequency and volume of uncontrolled discharge from urban areas and diffuse pollution. Further, where surface drainage enters sewer systems, water treatment facilities may fail to contain the high water volumes during extreme stormwater surges, resulting in the overflow of untreated effluent (Wheater and Evans 2009). Although new developments cannot connect drainage from roofs directly to sewers, it remains applicable to a significant proportion of UK buildings. Sources of diffuse pollution are difficult to isolate. Significant uncertainty is associated with quantifying water pollution derived from movement through urban environments, with insufficient modelling capability cited by Wheater and Evans (2009) as a hindrance.

2.5 Summary

Wheater and Evans (2009) consider that the application of more 'radical' measures in the UK will become increasingly more plausible by 2050. They cite RWH, GW recycling and green roofs as suitable water management strategies for the future in order for supply to meet demand, especially in the south-east of the UK. Climate change is a key driver for maintaining supply in the south-east of England, but has, according to many WSPs, no predicted impact in other water resource zones (Wheater and Evans 2009). As a consequence, climate change is not given precedent as a requirement for investment by the water industry as a whole, although it is noted as a potentially key influence by some WSPs. With respect to flood mitigation, however, there is a substantially greater potential and one where impacts will be realised throughout the UK. A criticism of water management in the UK is over-reliance on a centralised supply system at the expense of small-scale local RWH and recycled GW approaches. In Australia, ecologically sustainable design includes water-sensitive urban design and use of RWH in the supplementation or overall substitution of reticulated urban water supply from a centralised supply facility (Mitchell 2004). Another benefit, of relevance to the UK and recent urban flooding, is that RWH is recognised within Australia for its potential to reduce the volume of stormwater discharge and peak run-off rate reduction (Coombes et al. 2002). Therefore the UK could potentially learn much from the Australian system of stormwater management, discussed in subsequent chapters. The UK Foresight Future Flooding project (Evans et al. 2004a, b) models scenarios based on both decreasing pressure on supply demand and the mitigation of flood risk, in combination. The increase in UK flood levels for the year 2080 predicted by the report has the potential to be reduced to current risk levels if suitable stormwater management and mitigation measures in combination were to be implemented.

References

Abdellatif M, Atherton W, Alkhaddar R et al (2015) Flood risk assessment for urban water system in a changing climate using artificial neural network. Nat Hazards 79:1059–1077

Ainger C, Butler D, Caffor I et al (2009) A low carbon water industry in 2050, Report: SC070010/R3. Environment Agency, Bristol

Alberto Campisano, David Butler, Sarah Ward, Matthew J. Burns, Eran Friedler, Kathy DeBusk, Lloyd N. Fisher-Jeffes, Enedir Ghisi, Ataur Rahman, Hiroaki Furumai, Mooyoung Han (2017) Urban rainwater harvesting systems: Research, implementation and future perspectives. Water Research 115:195–209

Arnell NW, Delaney EK (2006) Adapting to climate change: public water supply in England and Wales. Clim Chang 78:227–255

Battarbee R, Heathwaite L, Lane S et al (2012) Water policy in the UK: the challenges. RGS-IBG policy briefing. Royal Geographical Society (with IBG), UK

Beddington J (2008) Food, energy, water and the climate: a perfect storm of global events? Government Office for Science Briefing, UK

Berndtsson J (2004) Beneficial use of stormwater: a review of possibilities. Chalmers University of Technology, Gothenburg

Bowker P, Escarameia M, Tagg A (2007) Improving the flood performance of new buildings. Flood resilient construction. RIBA Publishing, London. Crown Copyright

British Geological Survey (2015) Groundwater for public supply. Available via http://wwwbgsacuk/research/groundwater/waterResources/GroundwaterInUK/2015html. Cited 10 June 2018

Brown RR, Sharp L, Ashley RM (2005) Implementation impediments to institutionalising the practice of sustainable urban water management. Presented at the 10th international conference on urban drainage, Copenhagen, August, pp 1–8

Building Regulations (2016) Approved Document G – sanitation, hot water safety and water efficiency (2015 edition with 2016 amendments). 84829. Crown Copyright, 2016, UK

Building Research Establishment Environmental Assessment Method BREEAM (2016) BREEAM In-use international. Technical manual. SD221–2.0:2015. February 2016. 83561 ©BRE Global Ltd 2016

Charlton MB, Arnell NW (2011) Adapting to climate change impacts on water resources in England – an assessment of draft water resources management plans. Glob Environ Chang 21:238–248

Climate Change Committee Adaptation Sub-Committee Progress Report (2012) Climate change – is the UK preparing for flooding and water scarcity. UK

Coombes PJ, Kuczera JD, Argue JR (2002) An evaluation of the benefits of source control measures at the regional scale. Urb Water 4(4):307–320

Curtis K, Cooper J (2009) Climate change toolkit. 07 Designing for flood risk. Second edition. Royal Institute of British Architects (RIBA) UK. ISBN 978-0-9561064-6-9

De Gouvello B, Gerolin A, Le Nouveau N (2014) Rainwater harvesting in urban areas: how can foreign experiences enhance the French approach? Water Sci Technol Water Supply 14(4):569–576

Department for Communities and Local Government (DCLG) (2008a) Code for sustainable homes: technical guide. Crown Copyright UK

Department for Communities and Local Government (DCLG) (2008b) Guidance on the permeable surfacing of front gardens. Crown Copyright UK

Department for Communities and Local Government (DCLG) (2009) Planning Policy Statement 25: Development and Flood Risk Practice Guide. Crown Copyright UK

Department for Communities and Local Government (DCLG) (2010) Code for sustainable homes: technical guide. November 2010. Crown Copyright UK

Department for Environment, Food and Rural Affairs (Defra) (2006) The development of the water industry in England and Wales. Crown Copyright UK

Department for Environment, Food and Rural Affairs (Defra) (2008) Future water strategy. London, UK

Department for Environment, Food and Rural Affairs (Defra) (2011) Water for life CM 8230. Crown copyright, UK

Department for Environment, Food and Rural Affairs (Defra) (2012) Distribution input and supply pipe leakage: 1992/93 to 2010/11. Available via http://wwwdefragovuk/statistics/environment/inland-water/iwfg14-drq/. Cited 21 July 2016

Department for Environment, Food and Rural Affairs (Defra) (2014) Water framework directive implementation in England and Wales: new and updated standards to protect the water environment. PB 14163. Crown copyright 2014

Department for Environment, Food and Rural Affairs (Defra) (2015) Development of zero and minimal herbicide regimes for controlling weeds on hard surfaces and determining their emissions. Project PS2802. Final report. Defra, UK

Department for Regional Development Northern Ireland (2016) Sustainable water – a long-term water strategy for Northern Ireland (2015 – 2040). Belfast. Available via https://www.infrastructure-ni.gov.uk/sites/default/files/publications/drd/sustainable-water-a-long-term-water-strategy-for-northern-ireland-2015-2040.PDF. Cited 21 July 2016

Diao K, Farmani R, Fu G, Astaraie-Imani M, Ward S, Butler D (2014) Clustering analysis of water distribution systems: identifying critical components and community impacts. Water Sci Technol 70(11):1764–1773

Dickie P (2006) Rich countries, poor water. World Wide Fund (WWF) Freshwater program, Zeist

Domènech L, Saurí D (2011) A comparative appraisal of the use of rainwater harvesting in single and multi-family buildings of the Metropolitan Area of Barcelona (Spain): social experience, drinking water savings and economic costs. J Clean Prod 19(6–7):598–608

Environment Agency (2009) Water for people and the environment – the water resources strategy for England and Wales. Environment Agency, Bristol, UK

Environment Agency (2010) Harvesting rainwater for domestic uses: an information guide. ea/br/e/std/v1. Environment Agency, Bristol, UK

Environment Agency (2013a) Groundwater protection: principles and practice (GP3). Version 1.1 August 2013. Environment Agency, Bristol, UK

Environment Agency (2013b) Rainfall runoff management for developments. SC030219. Environment Agency, Bristol, UK

Environment Agency (2015) Water supply and resilience and infrastructure. Environment Agency advice to Defra. October 2015. UK

Environment Agency (2018) Learn more about flood risk. Available via https://flood-warning-information.service.gov.uk/long-term-flood-risk/map. Cited 10 June 2018

European Commission (1998) Drinking water directive (Council Directive 98/83/EC of 3 November 1998 on the quality of water intended for human consumption). Brussels, Belgium

European Commission (2015) Drinking water directive amended (EU 2015/1787). Brussels, Belgium

European Union (EU) (2000) The EU water framework directive (WFD) – integrated river basin management for Europe. Brussels, Belgium

Evans EP, Ashley R, Hall JW et al (2004a) Foresight future flooding, scientific summary: Volume 1: future risks and their drivers. Office of Science and Technology, London. Available via http://www.foresight.gov.uk/OurWork/CompletedProjects/Flood/index.asp. Cited 21 July 2016

Evans EP, Ashley R, Hall JW et al (2004b) Foresight future flooding, Scientific summary: Volume 2: managing future risks. Office of Science and Technology, London

FAO (Food and Agriculture Organization of the United Nations) (2004) Capacity development in irrigation and drainage. Issues, challenges and the way ahead. FAO Water Report 26. In: Proceedings of the International Workshop held on 16 September 2003 during the International Commission on Irrigation and Drainage. Fifty-fourth International Executive Council Meeting. Montpellier, France

Fewkes A (2000) Modelling the performance of rainwater collection systems: towards a generalised approach. Urb Water 1(4):323–333

Fewkes A (2012) A review of rainwater harvesting in the UK. Struct Surv 30(2):174–194

Flood and Water Management Act (2010) HMSO, London

Gassert F, Reig P, Luo T et al (2013) Aqueduct country and river basin rankings: a weighted aggregation of spatially distinct hydrological indicators. Working paper. Washington, DC: World Resources Institute, November 2013. Available online at wri.org/publication/aqueduct-country-river-basin-rankings

Gedge D, Grant G, Kadas G (2009) Creating green roofs for invertebrates. A best practice guide. Buglife – The Invertebrate Conservation Trust, Peterborough

Godskesen B, Hauschild M, Rygaard M et al (2013) Life-cycle and freshwater withdrawal impact assessment of water supply technologies. Water Res 47(7):2363–2374
Griggs JC, Shouler MC, Hall J (1997) Water conservation and the built environment. In: Roaf S (ed) Architectural digest for the 21st century. School of Architecture, Oxford Brookes University, Oxford, pp 3–14
Growing Blue (2016) Growing blue: water. Economics. Life. Final report. Available via http://growingblue.com/the-growing-blue-tool/#. Cited 3 March 2016
Hassell C (2005) Rainwater harvesting in the UK – a solution to increasing water shortages? Proceedings of the 9th International Conference on Rainwater Catchment Cistern Systems. Petrolina, Brazil
HM Government (2014) National planning policy framework for England and planning practice guidance. Flood Risk and Coastal Change. Available via http://planningguidance.communities.gov.uk/blog/guidance/flood-risk-and-coastal-change/. Cited 21 June 2016
HM Government's Sustainable Development Strategy (2005) Securing the future. Cm 6467. Crown Copyright, UK
Hunt DVL, Rogers CDF (2014) A benchmarking system for domestic water use. Sustainability 6:2993–3018
Iveroth SP, Johansson S, Brandt N (2013) The potential of the infrastructural system of Hammarby Sjostad in Stockholm, Sweden. Energy Policy 59:716–726
Jones P, MacDonald N (2007) Making space for unruly water: sustainable drainage systems and the disciplining of surface runoff. Geoforum 38(3):534–544
Kallis G, Butler D (2001) The EU water framework directive: measures and implications. Water Policy 3:125–142
Kasiviswanathan KS, Sudheer KP (2017) Methods used for quantifying the prediction uncertainty of artificial neural network based hydrologic models. Stoch Env Res Risk A 31(7):1659–1670
Konig KW (1999) Rainwater in cities : a note on ecology and practice. In: Cities and the environment : new approaches for eco-societies. United Nations University Press, New York
Lang T, Schoen V (2016) Food, the UK and the EU: Brexit or Bremain? 8th March 2016. Food Research Collaboration Policy Brief
Li Z, Boyle F, Reynolds A (2010) Rainwater harvesting and greywater treatment systems for domestic application in Ireland. Desalination 260:1–8
Liu S, Butler D, Memon FA, Makropoulos CK (2007) Exploring the water saving potential of domestic greywater reuse systems. In: 4TH IWA specialist conference on efficient use and management of urban water supply (Efficient 2007), Jeju Island, Korea, May vol 1, p 115–121
Maltby E (2012) Promoting natural water retention – an ecosystem approach. DG Environment Sci for Environ Policy 32
Melville-Shreeve P, Ward S, Butler D (2016a) Rainwater harvesting typologies for UK houses: a multicriteria analysis of system configurations. Water 8(4):129
Melville-Shreeve P, Cadwalader O, Eisenstein W et al (2016b) Rainwater harvesting for drought management and stormwater control in the San Francisco Bay Area. Fonctionnement global du système d'assainissement/Global performance of the drainage system-Réutilisation des eaux pluviales/Stormwater reuse
Melville-Shreeve P, Ward S, Butler D (2016c) Dual-purpose rainwater harvesting system design. Sustainable surface water management: a handbook for SUDS 205
Met Office UK (2016) Map of UK rainfall average (1981–2010). Crown Copyright. Available via http://www.metoffice.gov.uk/learning/rain/how-much-does-it-rain-in-the-uk. Cited 20 July 2016
Mitchell VG (2004) Integrated urban water management: a review of current Australian practice, CMIT-2004-075, AWA/CSIRO, Canberra
O'Brien DC (2014) Sustainable drainage system (SuDS) ponds in inverness, UK and the favourable conservation status of amphibians. Urb Ecosyst 18:321–331
Palla A, Gnecco I, La Barbera P (2017) The impact of domestic rainwater harvesting systems in storm water runoff mitigation at the urban block scale. J Environ Manag 191:297–305

Parsons D, Goodhew S, Fewkes A et al (2010) The perceived barriers to the inclusion of rainwater harvesting systems by UK house building companies. Urb Water 7(4):257–265
Pericli A, Jenkins J (2015) Review of current knowledge: smart meters and domestic water usage. Marlow foundation for water research. Commissioned report
Pfister S, Koehler A, Hellweg S (2009) Assessing the environmental impacts of freshwater consumption in LCA. Environ Sci Technol (American Chemical Society) 43:4098–4104
Pitt M (2008) The Pitt review. Learning the lessons from the 2007 Floods. Cabinet Office, London, 505 pp
Schuetze T (2013) Rainwater harvesting and management – policy and regulations in Germany. Water Sci Technol Water Supply 13(2):376–385
Scottish Water (2010) Water Efficiency Plan 2011-2015. Available via http://www.scottishwater.co.uk/assets/domestic/files/you%20and%20your%20home/water%20efficiency/swwaterefficiencyplan.pdf. Cited 21 July 2016
Stuart ME, Manamsa K, Talbot JC, Crane EJ (2011) Emerging contaminants in groundwater. British geological survey open report, OR/11/013
United Kingdom Meteorological Office (2018) Available via https://www.metoffice.gov.uk/public/weather/climate/#?tab=climateMaps. Cited 10 June 2018
Ward SL, Butler D, Memon FA (2011) Benchmarking energy consumption and CO2 emissions from rainwater harvesting systems: an improved method by proxy. Water Environ J 26(2):184–190
Ward SL, Barr S, Butler D et al (2012) Rainwater harvesting in the UK – socio-technical theory and practice. Technol Forecast Social Change 79(7):1354–1361
Ward SL, Barr S, Memon FA et al (2013) Rainwater harvesting in the UK: exploring water-user perceptions. Urb Water J 10(2):112–126
Water Act (2003) HMSO, London, UK
Water Industry Act (1999) HMSO, London, UK
Water Supply Regulations (1999) HMSO, London, UK
Water UK (2016) Water resources long-term planning framework (2015–2065). Final Report
Waterwise (2012) Water – The Facts. Why do we need to think about water? Waterwise, London. http://www.waterwise.org.uk/data/resources/25/Water_factsheet_2012.pdf. Cited 5 July 2016
Wentworth J (2012) Houses of parliament parliamentary office of science and technology. Postnote 419. Parliamentary copyright 2012
Whatmore S et al (2010) Understanding environmental knowledge controversies: the case of flood risk management: Full research report ESRC end of award report, RES-227-25-0018. ESRC, Swindon
Wheater H, Evans E (2009) Land use, water management and future flood risk. Land Use Policy 26S:S251–S264
Whitehead PG, Wilby RL, Battarbee W (2009) A review of the potential impacts of climate change on surface water quality. Hydrol Sci J 51(1):101–123
Whitehead PG, Crossman J, Balana BB et al (2013) A cost-effectiveness analysis of water security and water quality: impacts of climate and land-use change on the River Thames system. Phil Trans R Soc A 371:20120413
World Bank (2018) Available via https://datacatalog.worldbank.org/. Cited 10 June 2018
World Resources Institute (2013) Water Stress by Country. Available via https://www.wri.org/resources/charts-graphs/water-stress-country Cited 11 June 2018

Chapter 3
Urban Water Quality

Buddhi Wijesiri and Ashantha Goonetilleke

Abstract Anthropogenic activities which are common to urban areas generate a range of physical, chemical and biological pollutants which are subsequently incorporated in stormwater runoff, leading to the deterioration of receiving water environments. This poses risks to both human and ecosystem health including carcinogenic and neurological effects and the loss of aquatic biodiversity. Water environments are an essential asset for enhancing urban liveability. Significant research has been undertaken in relation to stormwater pollutant characterisation and pollutant processes, which forms the baseline knowledge for developing effective stormwater pollution mitigation strategies. The current practice of formulating strategies to improve stormwater quality relies on the fundamental understanding that pollutants accumulate on urban surfaces during dry weather periods and are subsequently washed-off during rainfall. However, there are significant gaps in the current knowledge base in relation to how pollutant load and composition could vary temporally and spatially, which is critical for understanding the dynamic nature of stormwater quality in urban catchments. This acts as a major constraint to informed decision-making in the context of designing effective stormwater pollution mitigation strategies. Moreover, climate change is a significant influential factor in relation to urban stormwater pollution. The predicted changes to dry and wet weather conditions would lead to changes to pollutant accumulation on urban surfaces, change pollutant characteristics and increase the likelihood of discharging shock loads of pollutants to receiving waters. Research is needed to understand the complex mechanisms underpinning pollutant processes and their influential factors and the role of climate change in order to enhance the well-being of urban communities.

Keywords Climate change · Stormwater pollutant processes · Stormwater quality · Stormwater pollutants · Stormwater reuse · Stormwater treatment

---

B. Wijesiri · A. Goonetilleke (✉)
School of Civil Engineering and Built Environment,
Queensland University of Technology (QUT), Brisbane, QLD, Australia
e-mail: a.goonetilleke@qut.edu.au

V. Jegatheesan et al. (eds.), *Urban Stormwater and Flood Management*,
Applied Environmental Science and Engineering for a Sustainable Future,
https://doi.org/10.1007/978-3-030-11818-1_3

3.1 Introduction

As at 2014, 89% and 82% of Australia's and the UK's population (compared to 54% global average) were living in urban areas, and this is projected to reach 93% and 89%, respectively, by 2050 (66% global average) (UNDESA 2014). The consequent spread of the built environment and associated anthropogenic activities result in significant impacts on the natural environment due to the generation of physical, chemical and biological pollutants leading to potential human and ecosystem health risks. Infiltration of stormwater is reduced as urbanisation replaces the vegetated/permeable landscape with impervious surfaces, resulting in increased stormwater runoff volumes. Increased flow volumes with greater uniformity of land slope leads to an increase in runoff velocity (Jacobson 2011; Marsalek et al. 2007; Miller et al. 2014). Consequently, flood risks are created due to changes to the hydrologic regime. In this context, research and investigation are key to formulating effective strategies to mitigate the negative impacts of urbanisation.

In the context of stormwater pollution mitigation, there are challenges that can be categorised as organisational, financial and technical. Organisational and financial challenges include the need for effective coordination between organisations such as the various levels of government, environmental protection authorities and natural resource management groups. The management of funding allocated has been identified as a significant challenge (ANZECC and ARMCANZ 2000). A range of issues resulting from rapid changes in stormwater quantity and quality are not adequately addressed, and this is a major technical challenge for developing effective pollution mitigation strategies (Liu et al. 2013). Uncertainty resulting from the spatial and temporal variations in stormwater quantity and quality and incomplete scientific knowledge of the processes drive these variations. This in turn results in uncertainty in planning and management decision-making for the development of effective pollution mitigation strategies (WWAP 2012).

In this chapter, the types and sources of stormwater pollutants are discussed together with stormwater pollutant processes, the impacts of stormwater pollutants on human and ecosystem health, impacts of stormwater quality on treatment and reuse and the predicated impacts of climate change.

3.2 Stormwater Pollution

3.2.1 Pollutant Types and Sources

During dry weather periods, a range of pollutants generated from natural and anthropogenic sources accumulate on urban surfaces. Pollutants accumulated on urban surfaces can be mobilised and entrained in stormwater runoff and commonly discharged into receiving water bodies through hydraulically efficient drainage systems. Increased flows with associated pollutants deteriorate receiving water quality

(Hvitved-Jacobsen et al. 2010; Meyer et al. 2005; Petrucci et al. 2014). The common stormwater pollutants include particulate matter, toxicants, organic matter, nutrients and microbial matter.

3.2.1.1 Particulate Matter

Soil erosion generates most of the solids found in stormwater (Hvitved-Jacobsen et al. 2010). Additionally, abrasion products from surface wear and automobile use contribute to the accumulation of particulate matter. A number of factors including source, land use, characteristics of the impervious surface, climate and traffic volume influence the load and composition of particulate matter (Gobel et al. 2007; Goonetilleke et al. 2017). In addition to increasing turbidity and sedimentation of receiving water bodies, particulate matter acts as a mobile substrate for the transport of other pollutants (Birch and Scollen 2003; Chiew et al. 1997; Dong and Lee 2009; Duong and Lee 2011; Gunawardana et al. 2013; Herngren et al. 2006; Murakami et al. 2005; Zhao et al. 2017).

3.2.1.2 Toxicants

These primarily include heavy metals and hydrocarbons, which can pose significant risks to human and ecosystem health (Brown and Peake 2006; Gobel et al. 2007; Herngren et al. 2005). The main source of heavy metals and hydrocarbons is vehicular traffic, where exhaust emissions, leakages from fuel and lubrication systems, tyre and brake wear and road surface wear are the primary contributors (Mummullage 2015; Pitt and Voorhees 2004). Heavy metals and hydrocarbons such as polycyclic aromatic hydrocarbons (PAHs) are found in association with particulate solids. The load and composition of toxicants accumulated on urban surfaces can be distinguished between different particle size fractions (Dong and Lee 2009; Duong and Lee 2009; Herngren et al. 2006; Lau and Stenstrom 2005; Murakami et al. 2004; Xiang et al. 2010; Zhou et al. 2005).

3.2.1.3 Organic Matter

Organic matter on urban surfaces can originate from vegetation debris and micropollutants emitted from combustion systems (Björklund and Li 2017; Gobel et al. 2007). Once discharged and deposited in receiving waters, organic matter is subject to decomposition through microbial, physical and chemical processes. Microbial decomposition may result in the depletion of dissolved oxygen in the water body and thereby pose risks to ecological health and species diversity in aquatic systems.

3.2.1.4 Nutrients

Similar to organic matter, the enrichment of nitrogen and phosphorous in urban waters can significantly reduce dissolved oxygen as a result of the occurrence of algal blooms, and in turn, decreasing aquatic species diversity. The main sources of nutrients are fertiliser application, industrial discharges, detergents, animal waste and sewerage system leakages (Heisler et al. 2008; Liu et al. 2017; O'Neil et al. 2012). Similar to toxicants, Miguntanna et al. (2010) found that different species of nutrients such as nitrates and phosphates are associated with different particle size ranges.

3.2.1.5 Microbial Matter

Stormwater runoff carries a range of microorganisms including pathogenic organisms which can cause a direct impact on human health. These microbial pollutants are derived mainly from sewer and on-site wastewater treatment system leakages and animal waste, which also contain supplementary matter for microbial growth such as fats/oils (Hathaway and Hunt 2011; McCarthy et al. 2012; Carroll et al. 2009). Further, particulate solids in stormwater runoff provide a host surface for microbial growth, influencing their propagation and abundance (NHMRC and NRMMC 2011).

3.2.2 Typical Pollutant Concentrations in Stormwater

Chapter 3 of the *Australian Runoff Quality* (Duncan 2005) provides data on typical concentrations of pollutants commonly present in stormwater from various urban runoff sources. A review conducted in relation to the design of stormwater harvesting systems under the Australian National Stormwater Recycling Guidelines focused on roof water and stormwater (National Resource Management Ministerial Council (NRMMC) et al. 2009). While that review did not consider specific land use, data provided (Table 3.1, reproduced below) are useful, detailing a range of pollutants including pathogens, heavy metals, nutrients, PAHs and physicochemical indicators that can be expected to be present in urban stormwater runoff.

3.2.3 Pollutant Processes

In the context of urban stormwater management, the design of pollution mitigation strategies involves the quantitative assessment of the underlying mechanisms of pollutant build-up and pollutant wash-off processes. The pollutants undergo many intermediate processes such as resuspension, aggregation and redeposition during

Table 3.1 Typical concentrations of known pollutants in untreated stormwater runoff

Contaminant	Unit	Mean	Standard deviation	Percentiles 5	25	50	75	95
Pathogens								
Campylobacter (bacteria)	/100 ml	3.31	1.97	1.00	1.93	2.89	4.21	7.02
Cryptosporidium (protozoa)	/10 L	176	211	12	52	112	222	546
Giardia (protozoa)		1.81	2.08	0.12	0.55	1.17	2.29	5.55
Bacteria Indicators								
Coliforms	/100 ml	97,665	170,197	3369	17,668	44,884	106,860	355,988
Clostridium perfringens		925	1016	103	315	614	1153	2748
E. coli		59,339	71,939	3835	17,203	37,511	74,564	184,382
Enterococci		13,792	10,928	1621	6043	11,229	18,586	34,465
Faecal coliforms		96,429	82,740	4694	20,440	44,168	87,235	215,568
Faecal streptococci		29,771	21,717	3829	13,991	25,212	40,317	70,894
Somatic coliphages		17,530	20,917	1154	5088	11,115	22,083	54,704
Heavy metals								
Aluminium	mg/L	0.19	0.60	0.49	0.78	1.07	1.47	2.29
Arsenic		0.009	0.001	0.006	0.008	0.009	0.009	0.011
Barium		0.028	0.005	0.021	0.025	0.028	0.031	0.038
Cadmium		0.0198	0.0242	0.0015	0.0061	0.0127	0.0248	0.0606
Chromium		0.009	0.005	0.002	0.005	0.008	0.011	0.017
Copper		0.055	0.047	0.012	0.025	0.041	0.068	0.141
Iron		2.842	1.246	1.126	1.956	2.674	3.540	5.100
Lead		0.073	0.048	0.017	0.040	0.063	0.095	0.162
Manganese		0.111	0.046	0.054	0.079	0.103	0.134	0.197
Mercury		0.218	0.105	0.080	0.143	0.201	0.273	0.411
Nickel		0.009	0.004	0.004	0.007	0.009	0.011	0.017
Zinc		0.293	0.153	0.080	0.183	0.272	0.379	0.570
Nutrients								
Oxidised nitrogen	mg/L	0.680	0.446	0.132	0.361	0.592	0.900	1.523
Total dissolved nitrogen		3.28	2.61	0.68	1.55	2.59	4.19	8.22
Total Kjeldahl nitrogen		2.84	4.14	0.60	0.95	1.59	3.04	8.82
Total organic nitrogen		0.623	0.828	0.160	0.233	0.367	0.669	1.874
Total nitrogen		3.09	2.33	0.62	1.52	2.51	4.00	7.46
Filtered reactive phosphorous		0.664	0.762	0.050	0.204	0.430	0.839	2.037
Total phosphorous		0.480	0.413	0.075	0.207	0.367	0.620	1.261

(continued)

Table 3.1 (continued)

Contaminant	Unit	Mean	Standard deviation	Percentiles 5	25	50	75	95
Hydrocarbons								
Polycyclic aromatic hydrocarbons	μg/L	0.262	0.306	0.017	0.078	0.168	0.331	0.811
Physicochemical indicators								
Ammonia	mg/L	1.135	1.187	0.102	0.394	0.793	1.464	3.281
Bicarbonate alkalinity as $CaCO_3$		35.21	3.36	29.99	32.887	35.04	37.37	40.97
Biochemical oxygen demand		54.28	45.58	6.56	22.87	42.53	72.03	140.77
Chemical oxygen demand		57.67	17.22	32.90	45.41	55.75	67.85	88.72
Chloride		11.40	1.05	9.75	10.67	11.35	12.08	13.20
Oil and grease		13.13	8.11	3.43	7.45	11.47	16.93	28.25
Sodium		10.63	2.82	6.58	8.62	10.31	12.29	15.72
Suspended solids		99.73	83.60	19.01	45.41	77.24	127.19	254.47
Total dissolved solids		139.60	17.30	112.89	127.44	138.54	150.58	169.60
Total organic carbon		16.90	3.33	11.99	14.54	16.60	18.92	22.80
Turbidity	NTU	50.93	40.46	7.98	23.21	40.74	66.78	127.79
pH	–	6.35	0.54	5.50	5.98	6.33	6.70	7.27

Note: Adapted from National Resource Management Ministerial Council (NRMMC) et al. (2009)

the overall build-up and wash-off processes, resulting in pollutant redistribution. The processes are illustrated conceptually in Fig. 3.1. While undergoing these processes, the particle-bound nature of these pollutants can influence pollutant mobility, reactivity and toxicity, which in turn influences stormwater quality and the degradation of receiving waters. Consequently, the underlying mechanisms of pollutant processes and their influential factors need to be understood in order to design effective pollution mitigation strategies.

3.2.3.1 Pollutant Build-Up

Particle-bound pollutants accumulate on urban surfaces during dry weather periods. The load of accumulated pollutants depends on the antecedent dry period, the rate of deposition and redistribution. Environmental and anthropogenic factors, including climate factors, traffic characteristics (e.g. traffic volume and congestion, speed and the nature of vehicle use) and land use, influence the depositional and redistributional processes. These factors are also often interrelated. For example, traffic characteristics are related to land use, with diesel-operated heavy-duty vehicle use being common in industrial and commercial areas, whereas petrol-operated light-duty vehicular activities are typical to residential areas (Goonetilleke et al. 2017).

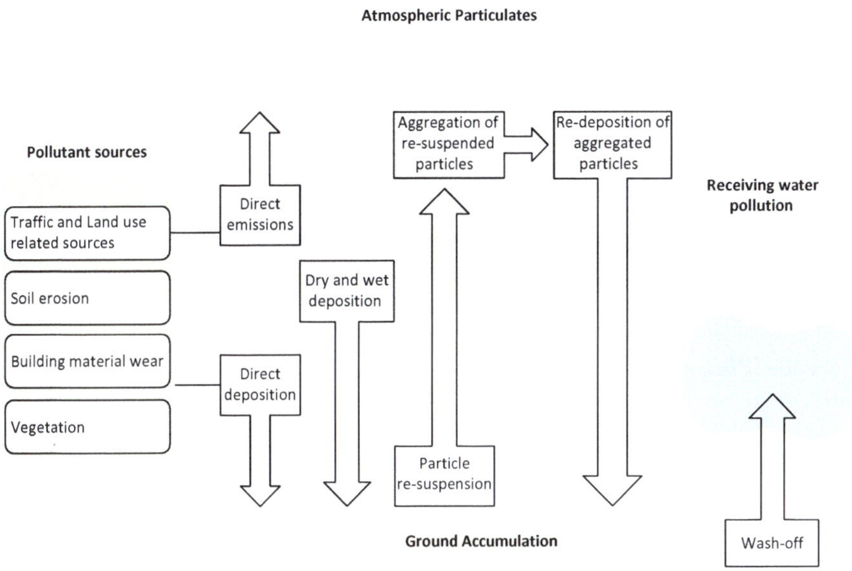

Fig. 3.1 Schematic diagram showing pollutant processes. (Adapted from Mummullage 2015)

The influential factors exert different impacts on the build-up of various pollutants. For example, land use significantly influences particulate matter build-up, while build-up of heavy metals and hydrocarbons such as PAHs is influenced by vehicle use (Mummullage et al. 2016). Moreover, the fact that the contribution of pollutants is a composite of a number of sources can also influence the build-up of stormwater pollutants. This means that pollutants generated from a particular source can interact with pollutants released from other sources over the antecedent dry period. For example, traffic-generated pollutants can have physical and chemical characteristics different from pollutants generated by industrial activities. Particulate matter originating from roadside soil and traffic have different particle size distributions, and different particle size fractions show different affinity to pollutants such as heavy metals and hydrocarbons (Jacobson 2011; Jayarathne et al. 2017; Marsalek et al. 2007). As pollutants interact, the size range of traffic-related particulates may change and, in turn, alter the affinity to other pollutants, thereby influencing the build-up of pollutants.

3.2.3.2 Pollutant Wash-Off

Pollutant wash-off is the process of mobilisation and transport of particle-bound pollutants accumulated on urban surfaces. The pollutants adhering to the impervious surface are detached by the impact of raindrops (kinetic energy) and runoff turbulence (Chiew et al. 1997; Egodawatta et al. 2007; Vaze and Chiew 2000). Although rainfall characteristics (intensity, duration, runoff volume and velocity) play a significant role in influencing the mechanisms of pollutant wash-off, the load of pollutants in wash-off is primarily influenced by the initial pollutant load available on the surface at the beginning of a storm event (Wijesiri et al. 2015a).

Pollutant wash-off can be defined by two concepts referred to as source limiting and transport limiting – the former being based on the amount of pollutants available on the surface and the latter representing the capacity of stormwater runoff to remove pollutants via wash-off (Fig. 3.2). The source-limiting concept states that almost all the pollutants accumulated are washed-off during a storm event, and pollutants will build up from zero over the following dry period. This concept is not valid for all storm events, because most events have the capacity to wash-off only a fraction of the initially available pollutant load due to the transport-limiting phenomenon (Zhao et al. 2016). The two concepts can be further differentiated by considering the wash-off behaviour of fine and coarse particles.

Zhao et al. (2016) noted that finer particles are likely to undergo a source-limiting process, which occurs at the initial period of a storm event (i.e. the phenomenon known as first flush) (Alias et al. 2014; Lee et al. 2002), while coarser particles

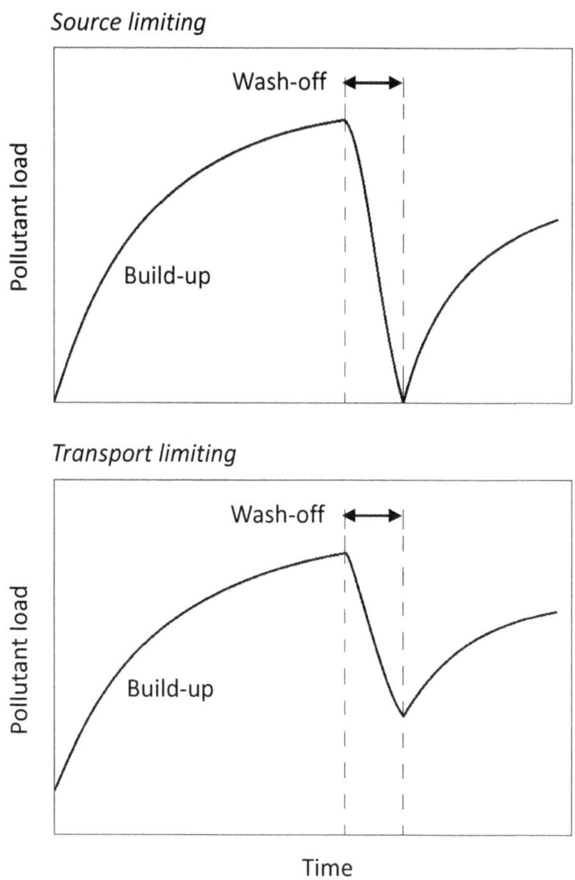

Fig. 3.2 Hypothetical representation of the source-limiting and transport-limiting concepts in relation to the pollutant wash-off process. (Adapted from Liu et al. 2015a)

would commonly undergo a transport-limiting process during the latter period of a storm event. On heavily trafficked roads, fine particles, which commonly carry a majority of particle-bound heavy metals, have been reported to accumulate in significantly higher loads compared to coarse particles. During wash-off of road-deposited particulate solids, the highest concentrations of heavy metals are found in association with the relatively finer particle fraction in stormwater runoff (Zhao et al. 2010). As such, pollutant wash-off processes, in general, can be considered as a combination of source- and transport-limiting processes.

3.2.3.3 Pollutant-Particulate Relationships and Mobility of Particle-Bound Pollutants

Strong relationships can be found between the mobility and the amount of particle-bound pollutants and how they are geochemically bound to particulate solids. For example, the geochemical description of heavy metals can be distinguished between those which are exchangeable, those associated with carbonates, those associated with Fe-Mn oxides and those associated with organic matter and residual fractions. Accordingly, the order of mobility of heavy metals has been distinguished as Cd > Zn > Pb > Co > Mn > Ni > Cu > Cr (Li et al. 2001; Manno et al. 2006; Tokalıoğlu and Kartal 2006). As such, heavy metals with higher mobility are likely to be weakly bound to particles. This is evident from the fact that Cd is associated with the carbonaceous fraction through weak electrostatic bonds, while Cu is bound to organic matter through strong covalent bonds (Banerjee 2003; Duong and Lee 2009).

Similar to heavy metals, different PAH concentrations and specific patterns of PAH distributions have been reported in different particle size fractions. This is not only evident during build-up, as PAHs also exhibit different distributions during wash-off (Murakami et al. 2004). For example, Nielsen et al. (2015) found that particles of size fraction <0.7 μm contain the highest concentrations of PAHs with four to six aromatic rings, which pose carcinogenic effects.

As such, these pollutant-particulate relationships influence the behaviour of particle-bound pollutants (Bae et al. 2002; Dong and Lee 2009; Lau and Stenstrom 2005). This can be related to pollutant adsorption by particles, which governs the relative mobility of dissolved and suspended pollutants in a solution (Bradl 2004; Sposito 2008).

3.2.3.4 Pollutant Adsorption by Particles

Stormwater pollutants are adsorbed by particulate matter through a mechanism known as surface complexation. This occurs when ionic and molecular forms of heavy metals and hydrocarbons interact with surface functional groups, which are chemically reactive molecular units protruding from a particle surface (Sparks 2003; Sposito 2008). The electrical charge developed on the particle surface by surface functional groups primarily influences surface complexation. Further, the chemical properties of surface functional groups (e.g. chemical structure and reactivity) specific to particulates of different sources diversify the particle surface charge distribution, which is influenced by particle physical and chemical characteristics (Barrow 2012; Zhang and Zhao 1997).

The surface charge density of a particle (charge per unit surface area) depends on particle size as the specific surface area (surface area per unit mass) increases with the decrease in particle size (Cristina et al. 2002; Gunawardana et al. 2012). This implies that the adsorption capacity of particulate solids deposited on urban surfaces varies as the particle size changes during build-up and wash-off processes.

The concentration of particle-bound toxic pollutants increases with the decrease in particle size. This is evident from the consistent variations in specific surface area and pollutant concentrations with particle size (Gunawardana et al. 2014; Lau and Stenstrom 2005). Accordingly, this pollutant-particulate relationship illustrates the differences in pollutant attachment to different particle size fractions, which in turn signifies how particle size influences variations in pollutant load and composition during pollutant build-up and wash-off processes (Gunawardana 2011; Wang et al. 2010).

Another particle characteristic that influences pollutant adsorption is surface coatings, such as hydrous metal oxides and organic matter. These coatings generate different electrical charge on the particle surface depending on the electrochemical properties of the surface (e.g. point of zero charge) and particle mineralogical composition (Barrow 2012; Zhang and Zhao 1997). Thus, inorganic oxides such as iron and aluminium oxides have the highest impact on varying both, positive and negative surface charge, while organic matter that produces negative charge specifically influences the adsorption of cationic forms of pollutants.

While several particle characteristics enhance pollutant adsorption, dissolved organic carbon (DOC) is reported to have suppression effects on adsorption, particularly during pollutant wash-off (Murakami et al. 2009). For example, elevated DOC levels in stormwater runoff are found to restrain the adsorption of the free heavy metal ions. This is due to the formation of stable metal-organic complexes (Förstner and Wittmann 2012; Naidu and Harter 1998). Similar to DOC, cations with different valency such as Na^+, Ca^{2+} and Zn^{2+} also exhibit suppression effects due to competition between cations for surface functional groups (Malamis and Katsou 2013; Valisko et al. 2007).

3.2.3.5 Variability and Uncertainty in Pollutant Processes

Pollutant build-up and wash-off processes are inherently uncertain due to their intrinsic variability that arises from the variation in pollutant load and composition over the antecedent dry period and the duration of a storm event. The knowledge of process uncertainty is critical as it influences planning and management decision-making in the context of designing effective stormwater pollution mitigation strategies. Hence, quantification of process uncertainty is essential, as its inherent nature constrains from being reduced or eliminated. This requires an in-depth understanding of the variability in pollutant processes (Kiureghian and Ditlevsen 2009; Wijesiri et al. 2016).

The behaviour of particles with different physical (e.g. size, density, surface charge distribution and mineralogy) and chemical (e.g. organic matter content and cation exchange capacity) characteristics is found to be different during build-up and wash-off (Gunawardana et al. 2013; Jain and Ram 1997). This results in the variations in load and composition of particle-bound pollutants. As such, particle behaviour primarily creates process variability and, in turn, process uncertainty (Badin et al. 2008; Mahbub et al. 2011; Vaze and Chiew 2004; Zafra et al. 2011).

During build-up, particles carrying pollutants initially deposit on urban surfaces. Subsequently, particles undergo redistribution due to influences such as vehicular traffic, street sweeping and wind. Moreover, particle characteristics are subject to change, resulting in changes to particle behaviour. For example, change in size due to aggregation of particles in the atmosphere can influence the rate of deposition (Kupiainen 2007; Sabin et al. 2006).

Particulate solids and associated pollutants built up on urban surfaces continuously resuspend in the atmosphere. Shear stress induced by tyres and turbulence created by vehicular traffic and wind have been identified as the primary driving forces in the resuspension of road-deposited particulate solids (Abu-Allaban et al. 2003; Thorpe and Harrison 2008).

Typically, light finer particles may be expected to be more easily resuspended than coarser dense particles. However, the laminar airflow that exists at the road surface due to vehicle movement is understood to prevent fine particles from resuspending (Hinds 2012; Mahbub et al. 2011; Patra et al. 2008). As such, the load and composition of pollutants resuspended in the atmosphere can vary depending on the pollutants associated with fine and coarse particle fractions, in addition to traffic characteristics that influence particle resuspension. The particles and associated traffic pollutants can also be resuspended during wash-off due to the effects of turbulent streams created by stormwater runoff (Wijesiri et al. 2015a; Zhao and Li 2013).

Once resuspended, particle behaviour can be distinguished between the period during resuspension and the period after resuspension. This is due to the change in particle characteristics, primarily particle size, during these two phases. Fine particles in suspension (atmospheric or stormwater runoff) tend to aggregate, forming coarser and heavier particles. Subsequently, aggregated particles redeposit and may undergo fragmentation due to the impact of raindrops and mechanical degradation (e.g. tyre abrasion). Further, change in particle characteristics during and after resuspension can result in changes in the load and composition of the pollutants associated with these particles.

3.3 Impacts of Stormwater Pollutants

3.3.1 Degradation of Physical and Chemical Quality of Water

The quality of urban receiving waters is influenced by the discharge of stormwater runoff transporting a wide range of pollutants. As such, irritation of eyes, skin and mucous membranes can occur due to accidental contact, ingestion or inhalation (Mallin et al. 2009; Tzoulas et al. 2007). Toxic pollutants associated with suspended solids can cause diseases in humans and affect species diversity in urban water ecosystems. Increase in suspended solids in receiving waters can also result in low dissolved oxygen levels, which can directly affect aquatic ecosystem health (Beach 2005).

3.3.2 Microbial Contamination of Water

Urban receiving waters can be contaminated by polluted stormwater due to pathogenic organisms from human and animal excreta. Different microorganisms cause a range of diseases depending on the concentration in water, virulence of the pathogen, per capita intake of water, infectious dose of a pathogen, risk of infection of a disease in a community and susceptibility of individuals. This in turn is related to the level of immunity and the demographics of the community (Jochimsen et al. 1998; NHMRC and NRMMC 2011).

3.3.3 Radiological Contamination of Water

Exposure to radiation even at low doses can potentially increase the occurrence of cancer and genetic disorders (Jirtle and Skinner 2007). Although, radiation exposure through water is minimal, urban waters can be contaminated by radionuclides where stormwater runoff originates from industrial areas and where there are naturally occurring radioactive elements in soil. In such cases, long-term effects on human health through drinking water and direct impact on aquatic ecosystems can be significant (ICRP 1999; Lokan 1998; Wrixon 2008).

3.4 Impacts of Stormwater Quality on Treatment and Reuse

Stormwater pollutants, which have different physical and chemical characteristics, behave differently while undergoing pollutant processes. Therefore, specific stormwater treatment measures should be implemented in order to meet the water quality standards. For example, vegetated systems such as bioretention basins are recommended for the removal of dissolved nitrogen in stormwater runoff (Taylor et al. 2005). In fact, recent research has highlighted the need for designing a range of measures targeting different species of stormwater pollutants (e.g. NO_x) for more effective treatment (Lucke et al. 2018).

Moreover, as discussed above, physical, chemical, microbial and radiological properties of stormwater can change depending on the type of pollutants entrained in runoff, thus varying the risks to human and ecosystem health. This will influence stormwater treatment for reuse, because different purposes (e.g. potable water, sanitation and recreational) require different levels of water quality (Fletcher et al. 2008; Liu et al. 2015b). As such, risk assessment in relation to the specific purpose of stormwater reuse is necessary to inform decision-making in the design of appropriate treatment systems.

3.5 Impacts of Climate Change on Stormwater Quality

The impact of climate change on the degradation of urban stormwater quality can be discussed in relation to three main aspects (Fig. 3.3): increase in the antecedent dry period between rainfall events, increase in intensity of typical rainfall events in a particular area and decrease in rainfall duration (AGO 2003; Delpla et al. 2009). The predicted increase in antecedent dry period can result in a range of impacts on

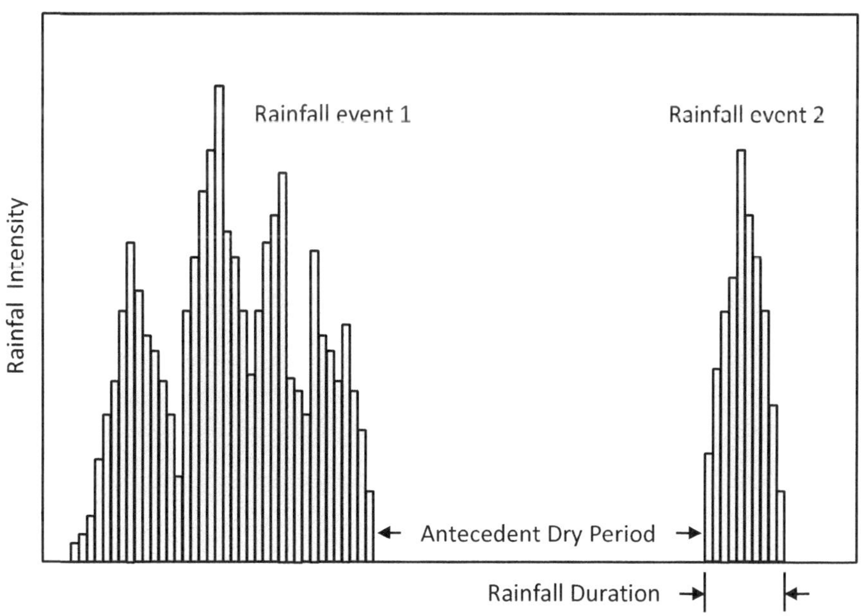

Fig. 3.3 Predicted changes in rainfall characteristics due to climate change

stormwater quality compared to changes to rainfall intensity and duration. Longer dry periods allow greater opportunity for the build-up of pollutants to occur on urban surfaces. However, the impact of natural and anthropogenic activities on pollutants accumulated on urban surfaces will also increase, resulting in further pollutant redistribution. Additionally, redistribution will change pollutant characteristics, particularly the characteristics of particulate solids (e.g. change in particle size due to tyre abrasion), which in turn will affect the adsorption of pollutants such as heavy metals and hydrocarbons.

Another impact of increased antecedent dry period is the potential for changes in chemical characteristics of pollutants due to photolysis and related reactions between pollutants including oxidation-reduction and hydrolysis. As the pollutants are exposed to different forms of light (e.g. infrared, visible, ultraviolet) over a longer period, the molecules will adsorb energy and can transform into different species. Moreover, the excited molecules (molecules with elevated energy) of a particular pollutant species can react with molecular forms of other pollutants and transform into different pollutant types (Miller and Olejnik 2001; Pirjola et al. 2012). These changes to pollutant characteristics due to exposure to the influence of natural and anthropogenic activities can result in changes to pollutant toxicity and mobility during build-up. The changes to pollutant mobility will influence the temporal variations in pollutant load and composition during build-up, which in turn will affect the wash-off of pollutants accumulated on urban surfaces during rainfall events (Wijesiri et al. 2015a, b).

Rainfall events with high intensity can increase the wash-off load of pollutants primarily due to the impact of the kinetic energy of raindrops. This increase in pollutant wash-off can be attributed to the predominant contribution of fine particles (caused by the first flush effect) to the total wash-off load in relatively shorter duration rainfall events. The increased first flush effect could overwhelm stormwater quality treatment devices such as Water Sensitive Urban Design (WSUD) measures. However, it is important to note that decreased duration of such rainfall events may reduce the total wash-off load due to the minimal contribution of coarser particles as these are mobilised primarily by the turbulent streams created by runoff (Zhao et al. 2016).

Given the significant climate change challenges (Stocker 2014), there is a critical need for strategic investment in solutions that deliver long-term sustainable outcomes. Across Australia and internationally, a growing body of urban water professionals are focussed on transitioning to more Sustainable Urban Water Management (SUWM) as they respond to the challenges associated with environmental degradation, rapidly growing urban population and the impacts from climate change (Novotny et al. 2010).

3.6 Knowledge Gaps

The review of current knowledge presented in this chapter identified future research directions that can potentially contribute to enhancing the knowledge base necessary to design effective stormwater pollution mitigation strategies and thereby safeguard urban waters. The current knowledge of pollutant processes does not adequately address the interactions between particles and particle-bound pollutants such as toxicants. These interactions potentially influence the variations in pollutant load and composition during pollutant processes, particularly during wet weather conditions, and in turn process variability and resulting process uncertainty. Therefore, the understanding of processes underpinning such interactions (e.g. adsorption and desorption by particles) needs to be improved. This will contribute to developing scientifically robust methods to accurately quantify pollutant process uncertainty, enabling informed decision-making.

3.7 Conclusions

Anthropogenic activities inherent to urbanisation influence environmental pollution such as the pollution of stormwater runoff that leads to the pollution of receiving waters. The consequences of stormwater pollution can be related to the degradation of the physical and chemical quality of water, microbial contamination of water and radiological contamination of water, posing risks to human and ecosystem health. Therefore, mitigating stormwater pollution is vital for safeguarding urban receiving waters, thereby improving urban liveability. However, stormwater pollution mitigation faces many organisational, financial and technical challenges.

The development of strategies to improve urban stormwater quality requires quantitative evaluation of pollutant processes. During dry weather periods, pollutants such as particulate matter, toxicants, microbial pollutants and nutrients undergo build-up on urban surfaces and subsequent wash-off and transport to receiving waters during storm events. The underlying mechanisms of pollutant processes are influenced by both anthropogenic and environmental factors. However, current knowledge lacks adequate understanding of these mechanisms to enable the development of effective pollution mitigation strategies.

Among several challenges in stormwater pollution mitigation, the impact of climate change is significant due to the predicted increase in the antecedent dry period between rainfall events, increase in rainfall intensity and decreases in rainfall duration. These phenomena can lead to the accumulation of large amounts of pollutants on urban surfaces, increase in the impact of natural and anthropogenic activities on pollutant processes, changes to pollutant characteristics and increase in the wash-off load of fine particles that carry a higher fraction of toxic pollutants.

References

Abu-Allaban M, Gillies JA, Gertler AW, Clayton R, Proffitt D (2003) Tailpipe, resuspended road dust, and brake-wear emission factors from on-road vehicles. Atmos Environ 37(37):5283–5293. https://doi.org/10.1016/j.atmosenv.2003.05.005

AGO (2003) Climate change: an Australian guide to the science and potential impacts. In: Pittock B (ed) Australia Australian Greenhouse Office, Canberra

Alias N, Liu A, Goonetilleke A, Egodawatta P (2014) Time as the critical factor in the investigation of the relationship between pollutant wash-off and rainfall characteristics. Ecol Eng 64:301–305. https://doi.org/10.1016/j.ecoleng.2014.01.008

ANZECC, and ARMCANZ (2000) Australian guidelines for urban stormwater management. Australian and New Zealand Environment and Conservation Council and Agriculture and Resource Management Council of Australia and New Zealand, Canberra

Badin A-L, Faure P, Bedell J-P, Delolme C (2008) Distribution of organic pollutants and natural organic matter in urban storm water sediments as a function of grain size. Sci Total Environ 403(1–3):178–187

Bae SY, Yi SM, Kim YP (2002) Temporal and spatial variations of the particle size distribution of PAHs and their dry deposition fluxes in Korea. Atmos Environ 36(35):5491–5500

Banerjee ADK (2003) Heavy metal levels and solid phase speciation in street dusts of Delhi, India. Environ Pollut 123(1):95–105

Barrow J (2012) Reactions with variable-charge soils, vol 31. Springer, Dordrecht

Beach J (2005) Safe drinking water. Lessons from recent outbreaks in affluent nations. Steve E. Hrudey and Elizabeth J. Hrudey. Published by International Water Association Publishing. ISBN: 1843390426. Price: £89.00. 486pp. Occup Med 55(8):643–644. https://doi.org/10.1093/occmed/kqi160

Birch GF, Scollen A (2003) Heavy metals in road dust, gully pots and parkland soils in a highly urbanised sub-catchment of Port Jackson, Australia. Soil Res 41(7):1329–1342

Björklund K, Li LY (2017) Adsorption of organic stormwater pollutants onto activated carbon from sewage sludge. J Environ Manag 197:490–497

Bradl HB (2004) Adsorption of heavy metal ions on soils and soils constituents. J Colloid Interface Sci 277(1):1–18

Brown JN, Peake BM (2006) Sources of heavy metals and polycyclic aromatic hydrocarbons in urban stormwater runoff. Sci Total Environ 359:145–155

Carroll SP, Dawes L, Hargreaves M, Goonetilleke A (2009) Faecal pollution source identification in an urbanising catchment using antibiotic resistance profiling, discriminant analysis and partial least squares regression. Water Res 43(5):1237–1246. https://doi.org/10.1016/j.watres.2008.12.017

Chiew FHS, Mudgway LB, Duncan HP, McMohon TA (1997) Urban stormwater pollution. Cooperative Research Centre for Catchment Hydrology, Clayton

Cristina C, Tramonte J, Sansalone J (2002) A granulometry-based selection methodology for separation of traffic-generated particles in urban highway snowmelt runoff. Water Air Soil Pollut 136(1–4):33–53. https://doi.org/10.1023/A:1015239831619

Delpla I, Jung AV, Baures E, Clement M, Thomas O (2009) Impacts of climate change on surface water quality in relation to drinking water production. Environ Int 35(8):1225–1233. https://doi.org/10.1016/j.envint.2009.07.001

Dong TTT, Lee B-K (2009) Characteristics, toxicity, and source apportionment of polycylic aromatic hydrocarbons (PAHs) in road dust of Ulsan, Korea. Chemosphere 74(9):1245–1253

Duncan HP (2005) Urban stormwater pollutant characteristics. In: Wong THF (ed) Australian runoff quality. Engineers Australia, Canberra

Duong TTT, Lee B-K (2009) Partitioning and mobility behavior of metals in road dusts from national-scale industrial areas in Korea. Atmos Environ 43(22–23):3502–3509

Duong TTT, Lee B-K (2011) Determining contamination level of heavy metals in road dust from busy traffic areas with different characteristics. J Environ Manag 92(3):554–562

Egodawatta P, Thomas E, Goonetilleke A (2007) Mathematical interpretation of pollutant wash-off from urban road surfaces using simulated rainfall. Water Res 41(13):3025–3031
Fletcher TD, Deletic A, Mitchell VG, Hatt BE (2008) Reuse of urban runoff in Australia: a review of recent advances and remaining challenges. J Environ Qual 37(5_Supplement):S-116–S-127. https://doi.org/10.2134/jeq2007.0411
Förstner U, Wittmann GT (2012) Metal pollution in the aquatic environment. Springer, Berlin
Goonetilleke A, Wijesiri B, Bandala ER (2017) Water and soil pollution implications of road traffic. In: Hester RE, Harrison RM (eds) Environmental impacts of road vehicles: past, present and future. The Royal Society of Chemistry, London, pp 86–106
Gunawardana CTK (2011) Influence of physical and chemical properties of solids on heavy metal adsorption. PhD, Queensland University of Technology
Gunawardana C, Goonetilleke A, Egodawatta P, Dawes L, Kokot S (2012) Role of solids in heavy metals build-up on urban road surfaces. J Environ Eng 138(4):490–498
Gunawardana C, Goonetilleke A, Egodawatta P (2013) Adsorption of heavy metals by road deposited solids. Water Sci Technol 67(11):2622–2629
Gunawardana C, Egodawatta P, Goonetilleke A (2014) Role of particle size and composition in metal adsorption by solids deposited on urban road surfaces. Environ Pollut 184:44–53. https://doi.org/10.1016/j.envpol.2013.08.010
Gobel P, Dierkes C, Coldewey WG (2007) Stormwater runoff concentration matrix for urban areas. J Contam Hydrol 91:26–42
Hathaway JM, Hunt WF (2011) Evaluation of first flush for indicator bacteria and total suspended solidsin urban stormwater runoff. Water Air Soil Pollut 217(1–4):135–147
Heisler J, Glibert PM, Burkholder JM, Anderson DM, Cochlan W, Dennison WC, Dortch Q, Gobler CJ, Heil CA, Humphries E, Lewitus A, Magnien R, Marshall HG, Sellner K, Stockwell DA, Stoecker DK, Suddleson M (2008) Eutrophication and harmful algal blooms: a scientific consensus. Harmful Algae 8(1):3–13. https://doi.org/10.1016/j.hal.2008.08.006
Herngren L, Goonetilleke A, Ayoko GA (2005) Understanding heavy metal and suspended solids relationships in urban stormwater using simulated rainfall. J Environ Manag 76:149–158
Herngren L, Goonetilleke A, Ayoko GA (2006) Analysis of heavy metals in road-deposited sediments. Anal Chim Acta 571(2):270–278
Hinds WC (2012) Aerosol technology: properties, behavior, and measurement of airborne particles. Wiley, Hoboken
Hvitved-Jacobsen T, Vollertsen J, Nielsen AH (2010) Urban and highway stormwater pollution: concepts and engineering. CRC Press/Taylor and Francis, London/New York
ICRP (1999) Protection of the public in situations of prolonged radiation exposure. ICRP Publication 82. Ann. ICRP 29(1–2)
Jacobson CR (2011) Identification and quantification of the hydrological impacts of imperviousness in urban catchments: a review. J Environ Manag 92:1438–1448
Jain CK, Ram D (1997) Adsorption of lead and zinc on bed sediments of the River Kali. Water Res 31(1):154–162
Jayarathne A, Egodawatta P, Ayoko GA, Goonetilleke A (2017) Geochemical phase and particle size relationships of metals in urban road dust. Environ Pollut 230:218–226. https://doi.org/10.1016/j.envpol.2017.06.059
Jirtle RL, Skinner MK (2007) Environmental epigenomics and disease susceptibility. Nat Rev Genet 8(4):253–262
Jochimsen EM, Carmichael WW, An J, Cardo DM, Cookson ST, Holmes CE, Antunes MB, de Melo Filho DA, Lyra TM, Barreto VST (1998) Liver failure and death after exposure to microcystins at a hemodialysis center in Brazil. N Engl J Med 338(13):873–878
Kiureghian AD, Ditlevsen O (2009) Aleatory or epistemic? Does it matter? Struct Saf 31(2):105–112
Kupiainen K (2007) Road dust from pavement wear and traction sanding, Monographs of the boreal environment research. Finnish Environmental Institute, Helsinki, p 26
Lau S-L, Stenstrom MK (2005) Metals and PAHs adsorbed to street particles. Water Res 39(17):4083–4092

Lee JH, Bang KW, Ketchum LH Jr, Choe JS, Yu MJ (2002) First flush analysis of urban storm runoff. Sci Total Environ 293(1–3):163–175. https://doi.org/10.1016/S0048-9697(02)00006-2

Li X, Poon C-S, Liu PS (2001) Heavy metal contamination of urban soils and street dusts in Hong Kong. Appl Geochem 16(11–12):1361–1368

Liu A, Egodawatta P, Guan Y, Goonetilleke A (2013) Influence of rainfall and catchment characteristics on urban stormwater quality. Sci Total Environ 444:255–262. https://doi.org/10.1016/j.scitotenv.2012.11.053

Liu A, Goonetilleke A, Egodawatta P (2015a) Role of rainfall and catchment characteristics on urban stormwater quality. Springer, Singapore

Liu A, Liu L, Li D, Guan Y (2015b) Characterizing heavy metal build-up on urban road surfaces: implication for stormwater reuse. Sci Total Environ 515:20–29. https://doi.org/10.1016/j.scitotenv.2015.02.026

Liu A, Carroll S, Dawes L, Goonetilleke A (2017) Monitoring of a mixed land use catchment for pollutant source characterisation. Environ Monit Assess 189(7):336. https://doi.org/10.1007/s10661-017-6047-7

Lokan K (1998) Drinking water quality in areas dependent on groundwater. Radiat Prot Aust 14(4):11–14

Lucke T, Drapper D, Hornbuckle A (2018) Urban stormwater characterisation and nitrogen composition from lot-scale catchments—new management implications. Sci Total Environ 619:65–71

Mahbub P, Ayoko GA, Goonetilleke A, Egodawatta P (2011) Analysis of the build-up of semi and non volatile organic compounds on urban roads. Water Res 45(9):2835–2844

Malamis S, Katsou E (2013) A review on zinc and nickel adsorption on natural and modified zeolite, bentonite and vermiculite: examination of process parameters, kinetics and isotherms. J Hazard Mater 252(Supplement C):428–461. https://doi.org/10.1016/j.jhazmat.2013.03.024

Mallin MA, Johnson VL, Ensign SH (2009) Comparative impacts of stormwater runoff on water quality of an urban, a suburban, and a rural stream. Environ Monit Assess 159(1):475–491

Manno E, Varrica D, Dongarrà G (2006) Metal distribution in road dust samples collected in an urban area close to a petrochemical plant at Gela, Sicily. Atmos Environ 40(30):5929–5941

Marsalek J, Cisneros BJ, Karamouz M, Malmquist P-A, Goldenfum JA, Chocat B (2007) Urban water cycle processes and interactions: urban water series-UNESCO-IHP, vol 2. Taylor & Francis, Boca Raton

McCarthy DT, Hathaway JM, Hunt, WF, Deletic A (2012) Intra-event variability of Escherichia coli and totalsuspended solids in urban stormwater runoff. Water Res 46(20):6661–6670

Meyer JL, Paul MJ, Taulbee WK (2005) Stream ecosystem function in urbanizing landscapes. J N Am Benthol Soc 24(3):602–612. https://doi.org/10.1899/04-021.1

Miguntanna NP, Goonetilleke A, Egodowatta P, Kokot S (2010) Understanding nutrient build-up on urban road surfaces. J Environ Sci 22(6):806–812. https://doi.org/10.1016/S1001-0742(09)60181-9

Miller JS, Olejnik D (2001) Photolysis of polycyclic aromatic hydrocarbons in water. Water Res 35(1):233–243

Miller JD, Kim H, Kjeldsen TR, Packman J, Grebby S, Dearden R (2014) Assessing the impact of urbanization on storm runoff in a peri-urban catchment using historical change in impervious cover. J Hydrol 515:59–70

Mummullage SWN (2015) Source characterisation of urban road surface pollutants for enhanced water quality predictions. PhD, Queensland University of Technology

Mummullage S, Egodawatta P, Ayoko GA, Goonetilleke A (2016) Sources of hydrocarbons in urban road dust: identification, quantification and prediction. Environ Pollut 216:80–85

Murakami M, Nakajima F, Furumai H (2004) Modelling of runoff behaviour of particle-bound polycyclic aromatic hydrocarbons (PAHs) from roads and roofs. Water Res 38(20):4475–4483. https://doi.org/10.1016/j.watres.2004.07.023

Murakami M, Nakajima F, Furumai H (2005) Size- and density-distributions and sources of polycyclic aromatic hydrocarbons in urban road dust. Chemosphere 61(6):783–791

Murakami M, Fujita M, Furumai H, Kasuga I, Kurisu F (2009) Sorption behavior of heavy metal species by soakaway sediment receiving urban road runoff from residential and heavily trafficked areas. J Hazard Mater 164(2–3):707–712
Naidu R, Harter RD (1998) Effect of different organic ligands on cadmium sorption by and extractability from soils. Soil Sci Soc Am J 62(3):644–650
National Resource Management Ministerial Council (NRMMC), Environmental Protection and Heritage Council (EPHC), National Health and Medical Research Council (NHMRC) (2009) Australian guidelines for water recycling: managing health and environmental risks (phase 2) – stormwater harvesting and reuse. National Water Quality Management Strategy, Canberra
NHMRC and NRMMC (2011) National water quality management strategy. National Health and Medical Research Council and Natural Resource Management Ministerial Council, Canberra
Nielsen K, Kalmykova Y, Strömvall A-M, Baun A, Eriksson E (2015) Particle phase distribution of polycyclic aromatic hydrocarbons in stormwater—using humic acid and iron nano-sized colloids as test particles. Sci Total Environ 532:103–111. https://doi.org/10.1016/j.scitotenv.2015.05.093
Novotny V, Ahern J, Brown P (2010) Water centric sustainable communities: planning, retrofitting and building the next urban environment. Wiley, Hoboken
O'Neil JM, Davis TW, Burford MA, Gobler CJ (2012) The rise of harmful cyanobacteria blooms: the potential roles of eutrophication and climate change. Harmful Algae 14(Supplement C):313–334. https://doi.org/10.1016/j.hal.2011.10.027
Patra A, Colvile R, Arnold S, Bowen E, Shallcross D, Martin D, Price C, Tate J, ApSimon H, Robins A (2008) On street observations of particulate matter movement and dispersion due to traffic on an urban road. Atmos Environ 42(17):3911–3926
Petrucci G, Gromaire M-C, Shorshani MF, Chebbo G (2014) Nonpoint source pollution of urban stormwater runoff: a methodology for source analysis. Environ Sci Pollut Res 21(17):10225–10242
Pirjola L, Lähde T, Niemi JV, Kousa A, Rönkkö T, Karjalainen P, Keskinen J, Frey A, Hillamo R (2012) Spatial and temporal characterization of traffic emissions in urban microenvironments with a mobile laboratory. Atmos Environ 63:156–167. https://doi.org/10.1016/j.atmosenv.2012.09.022
Pitt R, Voorhees J (2004) Sources of stormwater pollutants, including pollutant buildup and wash-off the source loading and management model (SLAMM)
Sabin LD, Hee Lim J, Teresa Venezia M, Winer AM, Schiff KC, Stolzenbach KD (2006) Dry Deposition and resuspension of particle-associated metals near a freeway in Los Angeles. Atmos Environ 40(39):7528–7538
Sparks, DL (2003) Environmental soil chemistry. Access Online via Elsevier
Sposito G (2008) The chemistry of soils. Oxford University Press, New York
Stocker T (2014) Climate change 2013: the physical science basis: Working Group I contribution to the Fifth assessment report of the Intergovernmental Panel on Climate Change. Cambridge University Press, Cambridge
Taylor GD, Fletcher TD, Wong TH, Breen PF, Duncan HP (2005) Nitrogen composition in urban runoff—implications for stormwater management. Water Res 39(10):1982–1989
Thorpe A, Harrison RM (2008) Sources and properties of non-exhaust particulate matter from road traffic: a review. Sci Total Environ 400(1):270–282. https://doi.org/10.1016/j.scitotenv.2008.06.007
Tokalıoğlu Ş, Kartal Ş (2006) Multivariate analysis of the data and speciation of heavy metals in street dust samples from the organized industrial district in Kayseri (Turkey). Atmos Environ 40(16):2797–2805
Tzoulas K, Korpela K, Venn S, Yli-Pelkonen V, Kaźmierczak A, Niemela J, James P (2007) Promoting ecosystem and human health in urban areas using Green Infrastructure: a literature review. Landsc Urban Plan 81(3):167–178. https://doi.org/10.1016/j.landurbplan.2007.02.001

UNDESA (2014) World urbanization prospects: the 2014 revisions, highlights. ST/ESA/SER.A/352. United Nations Department of Economic and Social Affairs, Population Division, New York

Valisko M, Boda D, Gillespie D (2007) Selective adsorption of ions with different diameter and valence at highly charged interfaces. J Phys Chem C 111(43):15575–15585

Vaze J, Chiew FHS (2000) A field study to investigate the effect of raindrop impact energy and overland flow shear stress on pollutant washoff. Proceedings of the 3rd international hydrology and water resources symposium (Hydro 2000), Perth, Australia

Vaze J, Chiew FH (2004) Nutrient loads associated with different sediment sizes in urban stormwater and surface pollutants. J Environ Eng 130(4):391–396

Wang C, Li Y, Liu J, Xiang L, Shi J, Yang Z (2010) Characteristics of PAHs adsorbed on street dust and the correlation with specific surface area and TOC. Environ Monit Assess 169(1–4):661–670. https://doi.org/10.1007/s10661-009-1205-1

Wijesiri B, Egodawatta P, McGree J, Goonetilleke A (2015a) Influence of pollutant build-up on variability in wash-off from urban road surfaces. Sci Total Environ 527–528:344–350. https://doi.org/10.1016/j.scitotenv.2015.04.093

Wijesiri B, Egodawatta P, McGree J, Goonetilleke A (2015b) Process variability of pollutant build-up on urban road surfaces. Sci Total Environ 518–519:434–440. https://doi.org/10.1016/j.scitotenv.2015.03.014

Wijesiri B, Egodawatta P, McGree J, Goonetilleke A (2016) Assessing uncertainty in pollutant build-up and wash-off processes. Environ Pollut 212:48–56. https://doi.org/10.1016/j.envpol.2016.01.051

Wrixon AD (2008) New ICRP recommendations. J Radiol Prot 28(2):161–168

WWAP (2012) The United Nations world water development report 4: managing water under uncertainty and risk, vol 1. UNESCO, Paris

Xiang L, Li Y, Yang Z, Shi J (2010) Influence of traffic conditions on polycyclic aromatic hydrocarbon abundance in street dust. J Environ Sci Health A 45(3):339–347

Zafra CA, Temprano J, Tejero I (2011) Distribution of the concentration of heavy metals associated with the sediment particles accumulated on road surfaces. Environ Technol 32(9):997–1008

Zhang XN, Zhao AZ (1997) Surface charge. In: Yu TR (ed) Chemistry of variable charge soils. Oxford University Press, New York

Zhao H, Li X (2013) Understanding the relationship between heavy metals in road-deposited sediments and washoff particles in urban stormwater using simulated rainfall. J Hazard Mater 246–247:267–276. https://doi.org/10.1016/j.jhazmat.2012.12.035

Zhao H, Li X, Wang X, Tian D (2010) Grain size distribution of road-deposited sediment and its contribution to heavy metal pollution in urban runoff in Beijing, China. J Hazard Mater 183(1–3):203–210

Zhao H, Chen X, Hao S, Jiang Y, Zhao J, Zou C, Xie W (2016) Is the wash-off process of road-deposited sediment source limited or transport limited? Sci Total Environ 563–564:62–70. https://doi.org/10.1016/j.scitotenv.2016.04.123

Zhao B, Liu A, Wu G, Li D, Guan Y (2017) Characterization of heavy metal desorption from road-deposited sediment under acid rain scenarios. J Environ Sci 51:284–293

Zhou J, Wang T, Huang Y, Mao T, Zhong N (2005) Size distribution of polycyclic aromatic hydrocarbons in urban and suburban sites of Beijing, China. Chemosphere 61(6):792–799

Chapter 4
Sustainable Drainage Systems (SuDS) in the UK

Andrew Green

Abstract Surface water and combined sewerage systems are traditionally used to collect and transfer storm water in the UK but have several disadvantages compared to Sustainable Drainage Systems (or SuDS). These disadvantages include a limited ability to treat water quality and a lack of adaptability to change, for example, the expansion of urbanised areas and increased frequency and severity of storm events due to climate change. Consequently SuDS have many features that potentially make them attractive to developers and local authorities, and, as a result, there is now a considerable emphasis on supporting the uptake of SuDS technologies in UK policy and legislation. However, a lack of commitment to the long-term delivery of SuDS is cited as a hindrance to more wide-scale uptake, coupled with an overarching sentiment that insufficient funds and other resources have been committed to flood resilience in the UK in general. Despite this, the number of potential component options that may be included in SuDS management trains in the UK is considerable, offering the identification and implementation of suitable combinations of options for a variety of situations. This chapter will identify and discuss these options, placing them in the context of current challenges to water supply and storm water management in the UK.

Keywords Attenuation tank · Bioretention system · Detention basin · Green roof · Infiltration trench · Pervious pavement · Soakaway · Swale · Urban flood mitigation · Sustainable drainage system · Water quantity management

---

A. Green (✉)
Agriculture and Environment Research Unit, University of Hertfordshire, Hatfield, UK
e-mail: a.green@herts.ac.uk

V. Jegatheesan et al. (eds.), *Urban Stormwater and Flood Management*,
Applied Environmental Science and Engineering for a Sustainable Future,
https://doi.org/10.1007/978-3-030-11818-1_4

4.1 Introduction

When rain falls on natural landscapes, it is generally either returned back to the atmosphere through evapotranspiration or it enters soil and groundwaters through infiltration (Woods Ballard et al. 2015). Some does eventually find its way into surface waters, but it generally does so over an extended period of time, since (in temperate environments at least) the proportion of precipitation transferred rapidly by overland flow is normally relatively low (Woods Ballard et al. 2015). Urban areas however, perturb the natural balance of hydrological systems in a number of ways (Woods Ballard et al. 2015; Miller et al. 2014; Barbosa et al. 2012). Firstly, they tend to have considerably less vegetative cover than their natural equivalents, reducing both the extent to which plants can pass water from the soil to the atmosphere through transpiration and the degree to which water is held up on vegetative and soil surfaces for subsequent loss by evaporation and/or infiltration (Woods Ballard et al. 2015). Secondly, urban catchments inevitably have a significantly higher proportion of impermeable surfaces, further reducing infiltration, and resulting in a considerable increase in both the magnitude and speed of surface runoff (Woods Ballard et al. 2015; Miller et al. 2014; Fletcher et al. 2013). As a result, catchment hydrology becomes more 'flashy', with shorter lag times and increased peak flows (Charlesworth et al. 2003; Miller et al. 2014). This in turn makes flooding more likely, particularly during intense rainfall events, either due to water being unable to enter the drainage network, being forced back out of it as the network becomes overwhelmed, or as a result of receiving surface waters overtopping their banks (Woods Ballard et al. 2015; Miller and Hutchins 2017). Increased flashiness also increases the likelihood of within-channel erosion of river beds and banks (Woods Ballard et al. 2015), which, as well as being a problem in its own right, may also change aquatic habitats, with a resulting impact on the natural biodiversity of the catchment.

Added to the problems associated with water quantity are a number of water quality issues (Woods Ballard et al. 2015; Miller and Hutchins 2017; Barbosa et al. 2012), with urban runoff often being heavily contaminated with a myriad of diffuse (non-point) source pollutants collected as it passes over impermeable surfaces. These include heavy metals, hydrocarbons, salts, particulates, organic matter and microbiological contaminants, to name but a few, all of which threaten the value of urban surface waters for biodiversity and as a human resource. In order to prevent blockages, our surface water sewerage systems are designed to flow at a rate sufficient to ensure that sediments remain suspended. This, however, means that both the sediments themselves and any contaminants associated with them (as well as those in solution) are rapidly transferred to receiving waters, often with only limited chance for biochemical degradation. Gully pots can provide a degree of storage for pollutants resulting from road runoff (particularly solid pollutants), which may permit some time for them to break down, but they can also become potent sources of pollution in their own right, with their efficacy having been shown to be heavily dependent on maintenance (i.e. the clearing of built-up sediments). In addition, in many older drainage systems, surface water is mixed with foul water (sewage) in so-called combined sewerage systems. As a result, high levels of surface runoff can lead to sewage treatment works having to cope with a greater than ideal rate of delivery, with the potential to impact on the level of treatment received. More locally however, it may result in excess flow in the network being discharged untreated to surface waters through combined sewer overflows (CSOs). One of the design principles behind such systems is that at the point in time when a CSO discharges, the level of surface runoff in the system will result in the sewage component being highly diluted, thereby minimising the damage caused. In practice however, increases in urbanisation have often outstripped any improvements made to the capacity of sewerage networks, resulting in discharges happening both more frequently and at lower levels of dilution than may have been the case in the past. This occurs particularly in areas in which the initial sewerage system was built to cope with the then level of expected runoff at the cheapest possible price, using pipes of a size that limits the network's ability to absorb increased flows. Even in newer separate sewerage systems however, the surface runoff being discharged untreated to surface waters can be contaminated with sewage, as a result of the presence of misconnections (Miller and Hutchins 2017), locations in which foul water sewerage systems have (either accidentally or deliberately) been connected to the surface water system.

The above problems have in general become worse over time, as populations have both increased in number and become more urbanised in nature. The world's population is already in excess of 7½ billion and predicted to be more than 11 billion by 2100 (UN DESA – Population Division 2017), whilst the proportion of people living in urban areas is expected to be around 66% by 2050, up from only 30% a century earlier (UN DESA – Population Division 2015). At a national level, the UK's population increased by 15 million between 1950 and 2015 and is expected to do the same again by 2100 (UN DESA – Population Division 2017), of which 82% now lives in urban areas (UN DESA – Population Division 2015). Consequently, urban areas have become both larger and more densely populated, placing urban drainage systems under increasing stress. On top of this, the future impact of climate change has to be considered. In the UK, for example, predictions for the impact of climate change on precipitation suggest that although its overall magnitude may remain approximately the same (or even decrease a little), the frequency of intense rainfall events is expected to increase in both summer and winter (Jenkins et al. 2009). This has serious implications for the future functioning of drainage systems which have been designed to capture a high proportion of rainfall and transfer it away from the area as quickly as possible, since not only is it likely that they will be unable to accommodate the necessary volumes of water in the time available (leading to localised surface water flooding) but also that water levels in receiving rivers will threaten neighbouring built-up areas more frequently than in the past. Climate change also has implications for the availability of water for human and natural exploitation during sensitive periods, with UK summers expected to get both hotter and drier (increasing losses through evapotranspiration), such that drought becomes a more serious threat (Jenkins et al. 2009), as it already is elsewhere in the world. Despite this however, our drainage system is founded on the principle of getting rid of water as efficiently as possible, with often little consideration of how that water could be utilised. In the UK, this is a particular issue in the south east of England, which already has significant pressures on its available resources in the summer months, due to its high (and increasing) population, pressure from other users (e.g. agriculture and power generation – although some may be returned to the environment for subsequent reuse) and the need to protect the environment (i.e. maintain sufficient flows to protect biodiversity). The functioning of surface drainage systems also means that the recharge of subsurface (e.g. soil) waters can be significantly reduced, since impervious surfaces prevent water from infiltrating, with it instead being carried away by surface water sewers. This has a knock on effect on the base flows of many urban rivers, reducing them below what would be the case in a natural catchment, and putting aquatic biodiversity under considerable stress in periods of low flow.

Simply building ever larger drainage systems to cope with more flashy conditions is simply not sustainable, in no small part due to the enormous levels of expense that would be involved, although given the age of much of the UK's sewerage network, for example, ongoing improvements are required. For example, large attenuation tanks, sometimes associated with devices intended to remove solid pollutants (filtration screens, hydrodynamic separators, etc.), can be retrofitted to the drainage network, but this comes at a high price and can rarely be guaranteed to cope with even the largest of flows (i.e. overflows may still be needed). Even if it were possible, it wouldn't solve the problem of river flooding (indeed it may make it worse) or aid in managing water as a resource. Instead a more integrated approach is required, which both minimises the challenges faced and provides economically viable methods for dealing with them. It is now recognised that sustainability can only be promoted if urban drainage is carried out in as 'natural' a way as possible (Charlesworth et al. 2003), and as a result SuDS (Sustainable Drainage Systems, sometimes called Sustainable Urban Drainage Systems - SUDS or Water-Sensitive Urban Design -WSUD) have become an increasingly important element in many new and renewed urban developments and have also been retrofitted in pre-existing situations (such as those that form the bulk of the UK's urban area), in an attempt to manage both water quality and quantity in a sustainable, economically viable way. In addition however, the implementation of SuDS provides opportunities to provide an urban environment that is beneficial for both natural biodiversity and human populations, through the provision of new green space (something that is central to many such systems) and the linking of habitats (Woods Ballard et al. 2015). This results in what have been described in the UK's SuDS Manual (Woods Ballard et al. 2015) as the four pillars of SuDS (Fig. 4.1).

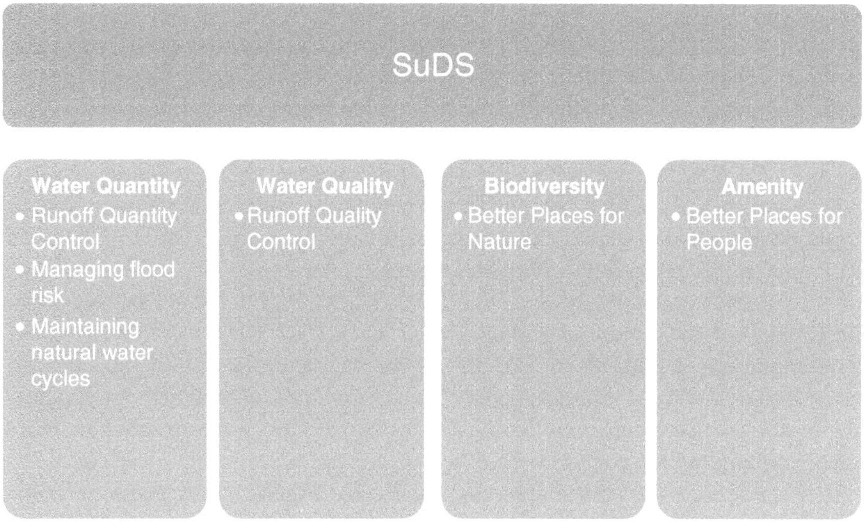

Fig. 4.1 The four pillars of SuDS. (Based on Woods Ballard et al. 2015)

Table 4.1 Classification of surface and groundwaters under the Water Framework Directive

Assessment class	Surface waters	Groundwaters
Ecological status (ecological potential if artificial or heavily modified)	Five classes – Based on biological and physico-chemical parameters	Not applicable
Chemical status	Two classes (good or fail) – Based on specific priority and other pollutants	Two classes (good or poor) – Based on core parameters and specific other pollutants
Quantitative status	Not applicable, although assessed in relation to ecological status	Two classes (good or poor)

European Parliament and the Council of the European Union (2000)

4.1.1 Relevant Legislation and Policies

Much of the UK legislation with relevance to SuDS currently has its basis in Europe, most notably in the form of the Water Framework Directive (WFD – European Parliament and the Council of the European Union 2000), which is the overarching legislation covering much of the European Union's water policy. It establishes the need to manage water at the catchment level and takes a holistic approach to the management of both water quantity and quality and as such has several implications for SuDS. In particular, the WFD sets quality standards for both surface and groundwaters, based on a combination of chemical, ecological and hydromorphological parameters (as detailed in Table 4.1), with the aim of ensuring that all waters reach at least good status. It also stresses the need to protect waters from non-point sources of pollution, including those from urban areas (European Parliament and the Council of the European Union 2000). In theory then, a surface waterbody is defined as being of good quality if its condition is close to that which might be found in a similar but unaffected waterbody of the same type. Although where waterbodies have been heavily modified, as is often the case in urban areas, this may not be ecologically possible, and so the concept of 'good ecological status' is replaced by one of 'good ecological potential', which requires a waterbody to get as close to good ecological status as possible within the constraints present.

Also of relevance at this scale is the Floods Directive (European Parliament and the Council of the European Union 2007), which establishes the framework for assessing and managing flood risks across the EU, with the aim of reducing the adverse consequences for human health, the environment, cultural heritage and economic activity that result from flooding. In as far as SuDS are concerned, member states are required to develop flood risk management plans for the river basin districts in their area of responsibility, and (in the UK at least) these plans stress the need for sustainable solutions to the problems associated with flood risk (European Parliament and the Council of the European Union 2007). In addition, the Urban Waste Water Treatment Directive (European Council 1991), the bulk of which deals with the collection and treatment of waste water rather than surface water, nevertheless requires that pollution from CSO discharges is limited, something to which the adoption of SuDS may contribute, by reducing the need for and frequency of overflows, although it may not necessarily reduce contaminant levels in those overflows that do occur.

Within the UK itself, the legislative and planning framework within which SuDS are developed and operated is complicated somewhat by the devolved nature of much of that legislation, as well as many of the decision-making powers associated with it, although, as detailed below, there are many similarities between countries. In England and Wales, the Flood and Water Management Act (particularly Schedule 3) required those with responsibility for managing flood risk (i.e. local authorities, etc.) to aim to make a contribution towards sustainable development, something which clearly has implications for the assessment and approval of SuDS. In particular, it required those bodies with responsibility for approving SuDS to grant that approval if they were satisfied that if constructed as proposed, it would comply with national standards for sustainable drainage and to refuse it if they were not satisfied that this would be the case. However, Schedule 3 of the Act was never fully enacted, instead, following a 2015 ministerial statement, the National Planning Policy Framework (DCLG 2012) was amended to make it clear that sustainable drainage systems should be provided for all major developments (as defined in the Town and Country Planning (Development Management Procedure) (England) Order 2015) unless it can be demonstrated to be inappropriate in the circumstances. It is also a requirement to ensure that SuDS are designed so as to take into account the likely future impacts of changes in climate and impermeable area over the expected life time of the system (DCLG 2012). However, much of the actual decision-making is carried out at a local level, so there can be considerable variation in the way in which these stipulations are applied, so when a SuDS might be considered appropriate for a given development can be spatially variable (DCLG 2012).

In Wales, planning policy is set out in Planning Policy Wales (Welsh Government 2016), which details the Welsh Government's commitment to sustainable development and stipulates that development proposals should:

> …include features that provide effective adaptation to, and resilience against, the current and predicted future effects of climate change, for example by incorporating green space to provide shading and sustainable drainage systems to reduce run-off….

This document is in turn supported by a series of Technical Advice Notes (TANs), chief amongst which in relation to SuDS is TAN 15 on Development and Flood Risk (National Assembly for Wales 2004) which, like the National Planning Policy Framework in England, encourages the use of SuDS. In particular, it makes it clear that new developments should not result in more runoff than was the case for the undeveloped site and that where possible, redevelopments should aim to reduce runoff. In addition, SuDS should be implemented "wherever they will be effective, in all new development proposals, irrespective of the zone in which they are located".

In Scotland, the central piece of legislation is the Water Environment and Water Services (WEWS) (Scotland) Act 2003, which, in as far as SuDS are concerned, requires the Scottish Government, the Scottish Environment Protection Agency (SEPA) and other responsible authorities to (where possible) "promote sustainable flood management" and (as in legislation south of the border) requires the use of SuDS in dealing with drainage from any new developments, a requirement which is reiterated in the Flood Risk Management (Scotland) Act 2009. Scotland's equivalents of the National Planning Policy Framework are the National Planning Framework (Scottish Government 2014a), which makes general mention of the need to ensure sustainable development and the Scottish Planning Policy (Scottish Government 2014b). This latter document, in particular, makes the following stipulations in relation to the planning system that have particular relevance to SuDS, namely, that it should (Scottish Government 2014b):

- Adopt a precautionary approach to flood risk, taking account of the predicted effects of climate change – as discussed elsewhere in this chapter, SuDS are considered more resilient to the impacts of climate change, although there are concerns regarding the impact it may have on the efficacy of some components.
- Reduce flood risk by (where appropriate) undertaking natural and structural flood management measures – which include the restoration of natural features and characteristics in the catchment and enhancing flood storage capacity, amongst other things.
- Avoid increased surface water flooding through requirements for SuDS and minimising the area of impermeable surface.

Scottish SuDS need to be designed in accordance with Scottish Water's specifications contained in the latest issue of the guidance document Sewers for Scotland (Scottish Water 2015), which, as well as setting out the planning requirements, contains many of the design principles also present in the SuDS Manual (Woods Ballard et al. 2015 – see Sect. 4.2), such as the need to manage runoff as near to source as possible, slow down runoff, treat it naturally and ensure that only good quality water is released to the wider environment. It also adopts the management train concept, incorporating the source, pathway and receptor hierarchy of management tools (i.e. management/treatment should occur as early in the management train as possible). Sewers for Scotland also stipulates the design requirements for both SuDS as a whole and a number of individual components (e.g. detention ponds – Scottish Water 2015).

Finally, Northern Ireland has its Strategic Planning Policy Statement (SPPS – Department of the Environment (Northern Ireland) 2015), which states that the planning system should help to mitigate and adapt to climate by (amongst other things):

> "...working with natural environmental processes, for example through promoting the development of green infrastructure and also the use of sustainable drainage systems (SuDS) to reduce flood risk and improve water quality."

It also urges planning authorities to encourage developers to use SuDS as the "preferred drainage solution", especially in areas susceptible to surface water flooding (Department of the Environment (Northern Ireland) 2015). There is also a specific Planning Policy Statement dealing with Planning and Flood Risk (Department of the Environment (Northern Ireland) 2014) which similarly encourages the uses of SuDS as the preferred form of drainage. In addition, the Northern Ireland Environment Agency (NIEA) has published Managing Stormwater: A Strategy for Promoting the Use of Sustainable Drainage Systems (SuDS) within Northern Ireland (Northern Ireland Environment Agency 2011), which is specifically intended to encourage the use of this form of drainage in a part of the UK generally lagging behind the mainland in terms of uptake (Department of the Environment (Northern Ireland) 2014). In 2016 the Water and Sewerage Services Act (Northern Ireland) 2016 further strengthened the push for greater use of SuDS by, for example, encouraging developers to consider the use of SuDS and giving Northern Ireland Water the power to refuse connections to the sewerage network if there are alternatives available (i.e. where they believe SuDS should be used instead).

4.2 Common SuDS Structures

In the UK, the most prominent source of guidance on the development of SuDS (although many others exist) is contained in the latest edition of the SuDS Manual (Woods Ballard et al. 2015), a "compendium of good practice" published by CIRIA with funding and support from many of the country's key government departments (e.g. Defra, DARD (now DAERA) and the Welsh Government), regulatory bodies (the Environment Agency, SEPA, NIEA) as well as industry leaders. This weighty document includes details of the underlying philosophy of SuDS, as well as their design, planning, construction and maintenance, from which it is clear that there are a bewildering number of components that could form part of an overall SuDS (not all of which can be covered here), some more or less suitable for different locations and/or issues to be addressed. Indeed, systems are generally comprised of a number of synergistic components, with each one feeding into another in order to provide holistic protection/treatment (the so-called SuDS management train – Woods Ballard et al. 2015). The overall approach to SuDS is to reinstate elements of what might be considered natural catchment functioning, by capturing rainfall, retarding its movement through the catchment, increasing losses through infiltration and evapotranspiration and using seminatural processes (through the action of vegetation and sunlight) to treat any remaining runoff. The central drivers in the development of a SuDS approach are therefore to (Woods Ballard et al. 2015):

- Use precipitation/surface waters as a resource – i.e. reducing demand on other resources (see storm water harvesting in Chap. 6).
- Manage rainfall as close to where it falls as practical – as opposed to what might be considered the traditional approach of transferring water elsewhere as fast as possible, for subsequent treatment and/or discharge.
- Manage runoff on the surface wherever possible – surface systems can generally cope with excess flows more easily than those underground, as where their design capacity is exceeded, it can often still be conveyed to a safe storage location. They also allow flood waters to rise gradually and visibly, such that local inhabitants can see potential problems developing and act accordingly.
- Allow as much rainfall to infiltrate as possible – although there may be situations in which this isn't appropriate, for example, where there is a danger of increasing the likelihood of groundwater flooding, many SuDS go some way towards increasing soil moisture content and, as a result, the natural base flows of urban watercourses.
- Promote evapotranspiration.
- Slow/store surface runoff – so as to attenuate flows and reduce peak discharges.
- Reduce pollution by a combination of source minimisation and runoff control – i.e. to reduce levels of pollution present in the catchment and delink pollutant sources from receiving waters.
- Treat runoff to minimise wider environmental implications.

To maximise the benefit of a SuDS then, a series of components should be combined in order to provide protection from and/or treatment of runoff at different stages of its journey from the point at which it lands to the receiving waters (and occasionally a little beyond). This 'SuDS management train' then relies on the combined functioning of the whole system, rather than any single component (see example in Fig. 4.2).

As discussed below, the final choice of options requires a clear definition of the objectives of a given SuDS and the site specific circumstances within which it operates, with each component being suitable for fulfilling one or more roles. Guidance documents contain matrices intended to guide the selection process (e.g. Woods Ballard et al. 2015; AECOM 2013), but the detailed selection and linking of options

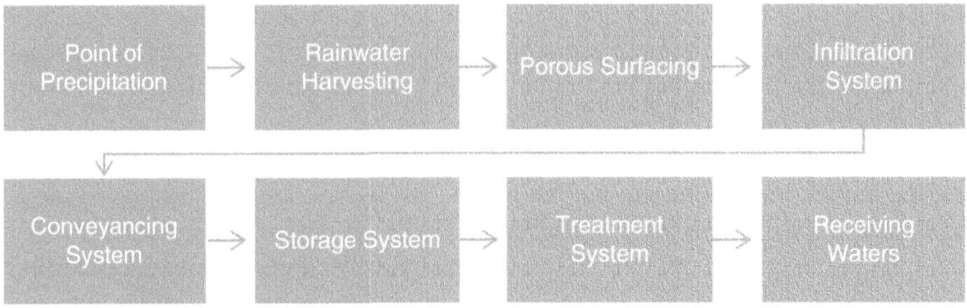

Fig. 4.2 Functioning of a SuDS based on the 'SuDS management train' approach. (Based on Woods Ballard et al. 2015)

requires a detailed understanding of their properties and the situation in which they will operate. Nevertheless, a selection of the major SuDS components and their benefits, as summarised from UK guidance in the SuDS Manual (Woods Ballard et al. 2015) supplemented by other guidance sources, is given in Sects. 4.2.1, 4.2.2 and 4.2.3.

4.2.1 SuDS Structures for the Streetscape

Pervious Pavements Pervious pavements are areas of hardstanding (car parks, pavements, roads, etc.), constructed from a range of materials, including block paving, gravels (sometimes bound in some way), concrete or plastic mesh reinforcing a grass surface and so on (Woods Ballard et al. 2015; US EPA 2005; Anglian Water Services 2011; Wilson et al. 2009). The UK's SuDS Manual divides pervious pavements into two categories, albeit that the aim of both is the same (Woods Ballard et al. 2015):

- *Porous pavements*: infiltrate water across their entire surface area (e.g. gravels, reinforced grass, porous concrete/asphalt)
- *Permeable pavements*: have a surface made of impermeable parts but with permeable joints (e.g. block paving)

The idea is to allow rainfall to percolate through the surface over a wide area, such that runoff from small rainfall events is avoided altogether. In larger rainfall events however, runoff will still occur such that some additional form of capture, treatment and/or transfer will be needed. In some forms of pervious pavement, underdrainage is added, so that some or all of the infiltrate is collected for onward transport, treatment and then use or discharge to the wider environment (Woods Ballard et al. 2015; US EPA 2005). All systems, however, provide benefits for both water quantity and quality control (through some combination of filtration, adsorption, biodegradation and sedimentation – Woods Ballard et al. 2015). Although as the bulk of pervious pavements encourage at least some infiltration, they should not be used in areas with relatively high levels of contamination in runoff (US EPA 2005).

Infiltration Trenches Infiltration trenches are a form of soakaway (see below), generally found alongside impervious areas of paving. They are in essence linear trenches, which (like nonlinear versions) are filled with gravel or similar material and which allow water to be stored for subsequent infiltration into the surrounding soil (Woods Ballard et al. 2015; US EPA 2005; Anglian Water Services 2011; Wilson et al. 2009). Consequently, infiltration trenches work best in areas with a reasonably high soil hydraulic conductivity, but such systems should not be used where runoff is expected to contain a high level of contamination, unless pretreatment is used, because of the risk posed to groundwater quality (US EPA 2005).

Swales In SuDS, swales (also known as grassed channels, dry swales, wet swales, biofilters or bioswales – US EPA 2005) are shallow, flat bottomed, vegetated (often grassed) channels, which both attenuate flow and treat water quality issues (Woods Ballard et al. 2015; Anglian Water Services 2011; Wilson et al. 2009). This is achieved through a combination of slowing runoff (by spreading it over the broader width associated with swales and increasing channel roughness through vegetation), the vegetative filtering and/or uptake of pollutants and by encouraging infiltration and evapotranspiration (Woods Ballard et al. 2015; Anglian Water Services 2011; Wilson et al. 2009). Further, some swales may be combined with check dams (or similar) to further retard flow at low levels of discharge and result in greater infiltration (EPA 2005; Anglian Water Services 2011; Woods Ballard et al. 2015). As linear features, they are well suited for dealing with runoff from roads and/or residential areas, in gently sloping catchments (US EPA 2005; Anglian Water Services 2011). Runoff should, however, not be excessively contaminated, as pollutant removal rates can be quite low (US EPA 2005), such that infiltrate may become a threat to groundwaters. In addition, since they take up a reasonably large area of land, they may not be suitable for very densely urbanised areas (US EPA 2005). As well as performing their water quantity and quality roles however, swales can also be an important amenity resource, increasing the level of urban greenspace (Anglian Water Services 2011; Wilson et al. 2009; Woods Ballard et al. 2015). Some can also be of considerable value to biodiversity, most notably wet swales (see below). The UK's SuDS Manual differentiates between three different classes of swale, namely (Woods Ballard et al. 2015):

- *Conveyance swales*: Basic, broad, vegetated channels which can perform all the roles mentioned above, although their biodiversity value may be limited unless planted with more than grass.
- *Dry (or enhanced) swales*: In this form of swale, the channel bed sits on top of an underdrained filtration bed, designed to increase both the capacity of the system (for flow attenuation) and the level of pollutant filtration achieved (for water quality improvement).
- *Wet swales*: This form of swale is specifically designed to have wet/marshy conditions at the bed (akin to a very gently sloping natural drainage system) with a mix of vegetation which is intended to increase the levels of treatment received (see wetlands below).

Filter Strips These are gently sloping areas of dense vegetation (e.g. grass) intended to receive runoff from neighbouring impermeable areas (roads, car parks, etc.) and remove sediments through a combination of sedimentation, filtration and vegetative uptake. In many cases they also reduce runoff volumes through increased infiltration, although this effect can be limited (EPA 2005; Anglian Water Services 2011; Wilson et al. 2009; Woods Ballard et al. 2015). As a result they can be beneficial for both water quantity and quality management. They work best in relatively flat areas, where runoff will enter them at low velocity and over a broad front, since concentration of flow tends to cause the strip to be overflowed too easily, and may result in much of the strip becoming redundant (Anglian Water Services 2011; EPA 2005).

Filter Drains Filter drains are linear features often to be found running along road edges, for example, and used for treating flow from impermeable surfaces (often after pretreatment using a filter strip or similar – Anglian Water Services 2011; Wilson et al. 2009; Woods Ballard et al. 2015). In essence they are a trench filled with gravel that allows runoff to be collected and transferred downstream more or less at the surface but with a built-in filtration ability which is intended to remove sediments and related pollutants, sometimes supported by adsorption and biodegradation processes. Some filter drains may also allow infiltration so as to reduce runoff volumes. Care is needed, however, to ensure that where rates of runoff are greater than the capacity of the filter drain, the excess can be stored/conveyed without resulting in flood damage (i.e. in a controlled way).

4.2.2 SuDS Structures for Open Spaces

Soakaways Simple soakaways have been around for some time, having been used to deal with runoff from domestic properties (e.g. roofs – Woods Ballard et al. 2015), particularly in more isolated areas without easy access to the sewerage network. In essence they are formed of a chamber (concrete or brick) filled with a material that will maintain its structure but also contain considerable void space in which water can be stored (e.g. rubble – Wilson et al. 2009; Woods Ballard et al. 2015). Stored water then infiltrates into the surrounding soil over an extended period of time. Larger, more modern systems however may use premade geocellular units (blocks that maximise void space) instead and may be combined with a siltation tank (Woods Ballard et al. 2015) to remove particulates prior to the water entering the soakaway, such that its water-holding capacity is maintained and infiltration isn't retarded. Such systems work best in areas where the soil has a reasonably high hydraulic conductivity (Wilson et al. 2009) but where the risk to underlying groundwaters is minimal (i.e. pollutant levels are not expected to be high and/or infiltration is prevented from getting into groundwaters by an impermeable layer, etc.).

Infiltration and Detention Basins These provide a similar service to soakaways but are surface-based systems comprised of a depression in which water is allowed to accumulate (a form of controlled flooding) and infiltrate into the soil, whilst contained sediments are deposited on the bed, together with any pollutants associated with them (Anglian Water Services 2011; Wilson et al. 2009; Woods Ballard et al. 2015). Where significant levels of sediment are expected however, some form of pre-basin settlement system is likely to be needed to prevent the basin silting up (Woods Ballard et al. 2015 Anglian Water Services 2011). As in the case of soakaways, infiltration basins work best on soils with reasonably high hydraulic conductivities and where there is no risk of polluting groundwaters (Anglian Water Services 2011). Detention basins are in many ways similar but play a slightly different role. Whilst infiltration basins are intended to reduce volumetric discharge (and peak flows) by allowing water to accumulate and then infiltrate, detention basins are intended to reduce peak flows (with a limited impact on volumetric discharge) by allowing water to accumulate and then be released to a downstream receptor (surface waters or a further SuDS component) at a reduced rate (Woods Ballard et al. 2015; Anglian Water Services 2011; Wilson et al. 2009). This is achieved by restricting the outflow discharge rate, for example, by forcing it to exit via a pipe of restricted size, resulting in water backing up into the detention area. Nevertheless, they may also be beneficial in relation to water quality due to the settlement of sediments and associated pollutants (sometimes followed by chemical/biological/physical breakdown and/or vegetative uptake – Woods Ballard et al. 2015). In addition to their water quantity/quality control roles however, both infiltration and detention basins can be useful as amenity resources (so long as significant contamination isn't expected) and can also be beneficial for biodiversity (Anglian Water Services 2011; Wilson et al. 2009).

Bioretention Systems/Areas Bioretention systems (sometimes called rain gardens, although these are generally smaller systems designed to treat runoff from a single property, e.g. Wilson et al. 2009; Woods Ballard et al. 2015) come in many different forms but are in essence shallow, vegetated depressions that can both reduce runoff volumes/rates and improve water quality (Woods Ballard et al. 2015; US EPA 2005; Anglian Water Services 2011; Wilson et al. 2009). Water quantity management is generally achieved by the interception of runoff, with subsequent losses through evapotranspiration and/or infiltration; they may also, however, attenuate that onward flow which does occur (Woods Ballard et al. 2015). They go some way past infiltration basins, however, in utilising specifically engineered soils and vegetation to enhance improvements in water quality (Woods Ballard et al. 2015; US EPA 2005; Anglian Water Services 2011; Wilson et al. 2009). They are particularly useful in dealing with relatively small runoff events (flow from larger events is often directed past such systems so as to prevent damage – Woods Ballard et al. 2015), during which runoff is allowed to collect within the basin and is then filtered through the vegetative root zone (with some uptake of pollutants) and soil layers, with cleaned water either being allowed to infiltrate into the subsoil or being collected by an underdrain for onward transfer (Woods Ballard et al. 2015; Anglian

Water Services 2011; Wilson et al. 2009; US EPA 2005). They are also capable of treating runoff carrying elevated levels of some contaminants, and small systems can be fitted into densely populated areas (they can be placed in many small spaces within the urban matrix – US EPA 2005). In addition to providing benefits in terms of both flood mitigation and water quality, they can also be of considerable amenity (mainly due to improvements in the aesthetic appearance of an area but also through urban cooling, for example) and biodiversity value (Woods Ballard et al. 2015).

Ponds and Wetlands It is recognised that natural wetland systems (ponds, marshes, etc.) provide a number of benefits in relation to the management of both water quantity and quality, and in SuDS these benefits can be harnessed by constructed surrogates (it isn't generally considered appropriate to divert flow into an existing wetland due to the damage that could be caused to sensitive habitats). Like their natural cousins, they can also be very valuable habitats for biodiversity and important amenity elements within the urban landscape (Wilson et al. 2009). At their simplest, a pond may be used to provide flood peak attenuation (and to some extent volume reduction, through increased evaporation and/or infiltration) in much the same way as a detention pond, with the main difference being that ponds are designed to contain a permanent pool of water, albeit of a temporally variable size (US EPA 2005). More complex constructed wetland systems, however, incorporate a series of treatment steps and are designed to take advantage of a range of physical, chemical and biological treatment mechanisms (e.g. sedimentation, adsorption, vegetative uptake, biofiltration and microbial decomposition – Ellis et al. 2003) within different environments. For example, they may include some combination of deep water, horizontal flow across a planted soil surface, surface horizontal flow, subsurface horizontal and subsurface vertical flow in macrophyte beds. Perhaps the most important factor influencing the efficacy of treatment in such systems is their hydraulic retention time (the average time that storm water remains in the wetland), with longer times generally resulting in greater pollutant removal (Ellis et al. 2003; Wilson et al. 2009). Such systems can be utilised across the UK (although in more arid parts of the world, the need for supplementary water to maintain a water content may be unjustified – US EPA 2005), albeit that the space required may not always be available in densely urbanised areas (US EPA 2005). They are often (but not always) placed at the lower end of a management train, so as to provide a final phase of water treatment before water is released to the wider environment (Woods Ballard et al. 2015; Wilson et al. 2009).

Trees Trees may seem a strange component to include within the definition of SuDS, but urban trees can in fact absorb considerable amounts of water, through interception, evapotranspiration and in some cases increased infiltration (Woods Ballard et al. 2015; Allen and Chapman 2001). A study carried out by the USDA, for example, showed that a medium-sized tree could intercept 2380 gallons ($\approx 9\ m^3$) of rain water a year (Geiger 2002), although the actual amount will be dependent on the tree species, rainfall intensity, temperature, wind speed and antecedent conditions, for example (Woods Ballard et al. 2015). There is also evidence that the overall impact of tree planting on the surface hydrology of a catchment is unlikely to be large unless the level of tree planting is also high, although localised impacts might still be significant (van Dijk and Keenan 2007). Nevertheless, trees can also absorb considerable water pollution through the vegetative uptake of contaminated waters; they have additional benefits for air quality as a result of filtering out airborne contaminants (Woods Ballard et al. 2015), are of considerable amenity (e.g. greenspace provision, urban cooling, etc.) and biodiversity value and can help to reduce climate change through carbon sequestration (Woods Ballard et al. 2015).

4.2.3 Other SuDS Structures

SuDS can also be comprised of a number of other components that don't fit easily into the categories above (in addition to systems for water harvesting, which are covered elsewhere in this volume), but which nevertheless may make a significant contribution in water quantity and/or quality terms.

Green Roofs These areas of vegetation are deliberately planted on the roofs of buildings and can result in a number of benefits both in terms of drainage control and the building itself (Newton et al. 2007). Green roofs have the ability to reduce runoff volumes/peaks considerably, by intercepting rainfall and storing it on vegetative surfaces and in the substrate provided for them, from where it is returned directly to the atmosphere through evapotranspiration (Newton et al. 2007). The extent to which this occurs is, however, highly dependent on the type of roof and vegetation, the intensity of the rainfall event and the antecedent conditions (Newton et al. 2007). Water quality may also be improved through a combination of biological, chemical and physical processes occurring within the substrate, vegetative uptake, and the filtering of airborne contaminants (Newton et al. 2007). In addition, however, they can be highly beneficial for biodiversity (particularly if planted with that role in mind) and, where there is access, can serve as an amenity resource (Newton et al. 2007). In addition, they can keep buildings warmer in winter and cooler in summer (reducing the need for energy use in air-conditioning systems) and reduce the urban heat island effect (Newton et al. 2007), with implications for the climate change resilience of our cities. Indeed, any of these properties may be the primary reason for installation. Such systems are nothing new, having been used in traditional architecture in areas such as Scandinavia for hundreds of years, but have increasingly become a feature of modern buildings over the past 50 years, with modern versions broadly falling into one of two categories (Woods Ballard et al. 2015):

- *Extensive green roofs*: These rely on a thin layer of substrate and are planted with hardy, slow-growing drought-resistant (for when the substrate dries) vegetation, which generally requires very little maintenance. Such systems are fairly lightweight, making them more suitable for retrofitting to existing buildings or for installation on roofs where access is limited.

- *Intensive green roofs*: These have a considerably deeper substrate layer and can support a much more structurally variable mix of vegetation (they are sometimes referred to as roof gardens, although the two terms aren't entirely synonymous). However, they also require greater maintenance and are therefore well suited to roofs with easy access.

Attenuation Tanks Subsurface attenuation tanks are intended to perform many of the same tasks (in as far as water quantity/quality are concerned at least) as above-ground detention and infiltration basins, in that they store runoff for release over an extended period of time (reducing peak flows) and in some cases can allow for some reduction in flow volumes through infiltration, if there is no risk of contamination of groundwaters. They can be constructed in a number of different ways, including through the use of oversize pipes, precast or poured-on-site concrete structures and glass-reinforced plastic tanks amongst others (Woods Ballard et al. 2015), and can be combined with a flow control device at the downstream end. As such they are, in the main, a SuDS component designed to address flood risks, rather than any of the other three SuDS pillars, although in some cases, the stored water can be of amenity value if utilised in a storm water harvesting programme (Woods Ballard et al. 2015). However, they generally result in little or no improvement in water quality when used alone, although they can be combined with other SuDS components to ensure that this requirement is met (Woods Ballard et al. 2015).

Proprietary Treatment Systems A number of (generally subsurface) products are available for the treatment of runoff prior to discharging it either to surface waters or into subsequent SuDS components (Woods Ballard et al. 2015). Such systems include hydrodynamic separators (removal of sediments and associated pollutants), filtration systems (solids), oil separators, treatment channels and so on (Woods Ballard et al. 2015), and although the extent to which they can be considered SuDS components is debatable, as many will discharge the polluted part of the outflow to a traditional sewerage system for downstream treatment (or need regular maintenance to remove collected material), they may in some cases be considered part of the management train as a whole (Woods Ballard et al. 2015). This is particularly the case if they are used to treat runoff so that it is of a sufficiently good quality for subsequent SuDS components to complete the treatment process.

4.3 Potential Impacts of Climate Change on SuDS Performance

The spread of urbanisation has resulted in an increase in flooding in urban areas, which has been exacerbated by the impacts of climate change, in particular the effect it has on the intensity and frequency of rainfall (Ashley et al. 2007; Tourbier and White 2007). Traditional subsurface drainage systems offer only limited resilience to climate change, due to their limited capacity, which, once exceeded, will almost inevitably lead to flooding somewhere. SuDS therefore are generally seen as an effective and more sustainable solution to the problem (Tourbier and White 2007). For example, systems which increase infiltration (e.g. pervious pavement) or store flood water (e.g. flood retention basins, etc.) can significantly reduce flash flooding and reduce the need for downstream flood protection (Tourbier and White 2007; Zhou 2014). Many elements within SuDS management trains may, however, have their effectiveness impacted by the changes in weather likely to be associated with climate change, because it has been found that the efficacy of SuDS components may be limited during extreme rainfall events (Zhou 2014). Infiltration-based systems, for example, have been shown to be most effective in relation to small, relatively frequent rainfall events (Holman-Dodds et al. 2003). Climate change, however, is likely to increase the intensity of rainfall events, such that a greater proportion of rainfall is likely to result in surface runoff. Similarly, retention-based systems may have to cope with a greater rate of discharge from the catchment. This may then result in exceeding their capacity to attenuate runoff, diminishing their capacity to prevent flooding, although the magnitude of any flooding may still be reduced. In addition, average retention times in components such as swales and wetlands may be significantly reduced, limiting the extent to which chemical and biological processes can capture/degrade pollutants (Nascimento et al. 1999).

Nevertheless, as the effects of climate change continue to take hold, not only will the ability of SuDS to cope with intense rainfall events become increasingly important but so will some of their other characteristics. The hard surfaces that are characteristic of built-up areas, heat up in hot weather and act as a heat store for the city as a whole, which (combined with reduced air movement due to the presence of buildings) causes urban temperatures to exceed those in surrounding rural areas. High urban temperatures have been shown to increase mortality in vulnerable human populations (e.g. the elderly), not least in locations such as the UK, where few domestic residences have the availability of air conditioning (Woods Ballard et al. 2015). The increased presence of vegetation and surface water that is associated with SuDS has the effect of cooling air temperatures (e.g. through shade and the cooling effects of evaporation), mitigating against the effects of the increased temperatures (particularly summer temperatures – Jenkins et al. 2009) associated with climate change.

4.4 SuDS Design Practice and Field Implementation

4.4.1 First Flush in SuDS Design

During storm events, a significant spike in the concentration of pollutants in runoff can occur on the rising limb of a flood hydrograph (Fig. 4.3), during which time the pollutant concentration being delivered to receiving waters can be considerable. This 'first flush' phenomenon is most commonly associated with those insoluble substances which accumulate on impermeable surfaces (e.g. roads, roofs, etc.) between rainfall events (e.g. sediments, heavy metals, hydrocarbons, pathogenic

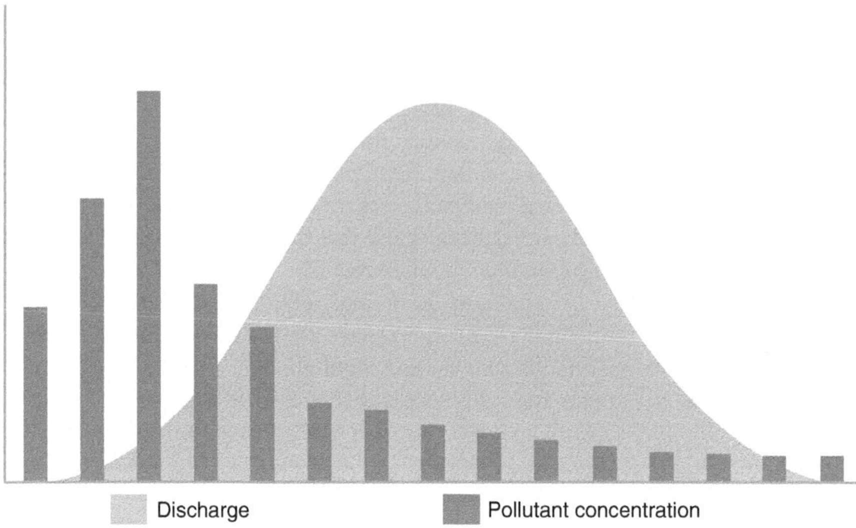

Fig. 4.3 First flush of pollutants during a hypothetical storm event

bacteria, etc.), and which are then mobilised in the early stages of runoff. In reality however, it can also affect soluble pollutants due both to the flush-off of those substances as rainfall begins and due to the flushing through of water stored in gully pots and elsewhere in the drainage network. For example, the dissolved oxygen content of runoff, which tends to be high in rainfall, can be very low during a first flush due to the remobilisation of anoxic waters. Estimates of the impact of a first flush in terms of total pollutant load vary considerably (City of Austin 1990; Bertrand-Krajewski et al. 1998) and are site, pollutant and event dependent (Lee et al. 2002). They are also heavily influenced by catchment area (being more clearly identifiable in smaller catchments), rainfall intensity, catchment properties (i.e. the proportion of impervious surface present, intensity of road traffic, etc.) and the length of the antecedent dry period (Lee et al. 2002). Regardless of the overall pollutant load significance however, high pollutant concentrations at a point in time when dilution may yet to have reached a maximum (i.e. flow rate is still increasing) have the potential to result in direct toxicity (acute effects) on aquatic biodiversity. This also has implications for the use of urban runoff as a water resource, since the early stages of a runoff event may be of too poor a quality for easy use (Chap. 6).

As a result of this, it is considered essential for all drainage systems, regardless of type, to be able to adequately treat first flush flows (Anglian Water Services 2011; Robert Bray Associates 2012), if the impact on receiving waters is to be minimised. In combined sewerage systems, it is generally hoped to achieve this by ensuring that the first flush component of a hydrograph can be accommodated within the network and passed on for subsequent treatment at a sewage treatment works. In separate sewerage systems however, this is not the case, since all surface water flow is passed on to local watercourses, such that a SuDS approach has the potential to result in significant benefits. To some extent this is achieved in a SuDS, by ensuring that small rainfall events (for which the first flush can be particularly damaging due to a lack of dilution in both the drainage network and receiving waters) do not result in discharge to receiving waters, for example, through the use of permeable pavement (Anglian Water Services 2011). However, for some contaminants, particularly those that don't tend to degrade, this can result in the build-up of contaminants within the catchment, for subsequent mobilisation at a later date, albeit that in a larger event dilution may mitigate the impact somewhat. For larger events, or runoff that cannot be treated in this way, first flush flows should be captured and sufficiently treated to ensure that discharge to receiving waters is of an acceptable quality (Anglian Water Services 2011). This may involve a set of treatment processes in series, such as those associated with filter strips, swales, bioretention areas, ponds and wetlands, for example (Anglian Water Services 2011; Woods Ballard et al. 2015 – see Sect. 3.2). It is important however that these components do not in themselves become sources of pollution for subsequent mobilisation, particularly in extreme rainfall events (although it is to be hoped that in such an event, levels of dilution would be significant). In wetlands, ponds and other related SuDS components, for example, pollutants that aren't readily broken down and/or absorbed by vegetation may simply accumulate and if not sufficiently immobilised could themselves be flushed. In some cases ongoing management, perhaps in the form of occasional dredging to remove polluted sediments, may be needed to prevent this, although as discussed below, SuDS components should be as self-sustaining as possible (Woods Ballard et al. 2015). It is therefore preferable for SuDS components to be sufficiently tailored to the expected pollutants (in as far as this can be known), so that they are capable of treating/removing them with minimal ongoing intervention.

4.4.2 Guidelines for Setting SuDS Design and Performance Objectives

Although many other national and local guides exist (e.g. AECOM 2013; Anglian Water Services 2011; Wilson et al. 2009; Essex County Council 2016, etc.), the main guidelines for SuDS development in the UK are contained in the SuDS Manual published by CIRIA (the Construction Industry Research and Information Association – Woods Ballard et al. 2015), and it is in the main these that are referred to in the remainder of Sect. 4.4. They establish the overarching principle of SuDS design to be to achieve "maximum benefit" in relation to each of the so-called four pillars of SuDS as shown in Fig. 4.1 (based on Woods Ballard et al. 2015), albeit that water quantity and quality are likely to be the main drivers behind most SuDS, and the extent to which each objective can be achieved will be site and circumstances specific. The water supply objective utilised in Australia is incorporated into the UK's water quantity objective, which is perhaps a reflection of the as yet lower emphasis being placed on this aspect of SuDS in the UK, due to lower levels of water stress (Gassert et al. 2013 – although as discussed above some areas of the UK do suffer from significant water stress). These same guidelines also establish the

Table 4.2 SuDS design criteria

Pillar	Design criteria
Water quantity	Use surface water as a resource
	Support the management of flood risk in the receiving catchment
	Protect morphology and ecology in receiving surface waters
	Preserve and protect natural hydrological systems on the site
	Drain the site effectively
	Manage on-site flood risk
	Design system flexibility/adaptability to cope with future change
Water quality	Support the management of water quality in the receiving surface and groundwaters
	Design system resilience to cope with future change
Amenity	Maximise multifunctionality
	Enhance visual character
	Deliver safe surface water management systems
	Support development resilience/adaptability to future change
	Maximise legibility
	Support community environmental learning
Biodiversity	Support and protect natural local habitats and species
	Contribute to the delivery of local biodiversity objectives
	Contribute to habitat connectivity
	Create diverse, self-sustaining and resilient ecosystems

Based on Woods Ballard et al. (2015)

design criteria that should be taken into account in attempting to achieve each of these objectives (as detailed in Table 4.2 – Woods Ballard et al. 2015), together with a number of cross-cutting criteria that will apply to all designs, namely, constructability, maintainability, cost-effectiveness and health and safety.

These of course are not mutually exclusive, with many SuDS components contributing to more than one objective (Woods Ballard et al. 2015). For example, a constructed wetland can simultaneously attenuate flood flows, remove pollutants, provide a habitat for biodiversity and improve the quality of the urban environment for those living and working there.

4.4.3 Targets for SuDS Implementation

Current UK guidelines for SuDS implementation (Woods Ballard et al. 2015) set reasonably detailed design criteria (objectives) and standards (performance targets) in relation to each of the four SuDS pillars, as summarised below.

Water Quantity SuDS are particularly useful in managing the flood risk associated with short, high-intensity rainfall events in relatively small catchments (Woods Ballard et al. 2015). Large rivers, in which flood risk is more likely to be associated with prolonged rainfall events over quite large areas (the impact of localised short-term events is often diminished when considered in terms of a major catchment), may benefit less from any individual scheme; however, that does not mean that large-scale implementation would not be beneficial, since the cumulative impact could be substantial (Woods Ballard et al. 2015). The targets established for water quantity relate to both controlling the rate at which water is discharged from an area and the total volume of water discharged, the goal being to promote as natural a flow regime as possible (Woods Ballard et al. 2015). As already discussed, urbanisation tends to increase both properties substantially, with implications for flooding, erosion and the maintenance of natural flow regimes and the ecosystems that depend on them. To this end, the UK's SuDS Manual establishes seven broad design criteria (Woods Ballard et al. 2015).

- Firstly, to use surface water as a resource, since this not only contributes to the conservation of water resources but also reduces the volume of water remaining to be discharged (this is discussed in greater detail in Chap. 6).
- Secondly, to manage flooding in the receiving catchment; and thirdly, to protect the natural morphological and ecological functioning of the receiving catchment. These related criteria are managed through the control of the volume of and rate at which water is discharged, together with a prioritisation of the ways in which water is discharged (Fig. 4.4 – note: discharge to foul sewers should not be considered and would not constitute a SuDS element). The volume element is controlled through the maximisation of use, infiltration and evapotranspiration, whilst peak flows are managed though flow attenuation (either by slowing water's general rate of movement through the catchment or capturing it for release over an extended period of time – the former generally being preferable).
- Fourthly, to preserve as much of the natural hydrological system (e.g. wetlands, streams, etc.) as possible and in so doing preserve as much of the natural functioning of the catchment as possible.
- Fifthly, to drain the site effectively, so as to reduce within catchment flooding. To achieve this, the system should drain sufficiently quickly to ensure (in as far as practical) that the ability of a system to store/pass the flow resulting from a rainfall event isn't reduced by flow from previous events still being present whilst not resulting in the catchment becoming excessively flashy. In other words, obtaining the correct gradient (within the constraints of morphology) can be something of a balancing act.
- Sixthly, to manage on-site flood risks, by ensuring that (in as far as practical – taking into account the sensitivity of the surrounding area) flood waters are contained within the conveyancing system and/or its associated storage systems.
- Seventhly, to ensure that the system is designed in such a way as to mean that it will continue to function appropriately as the catchment (e.g. levels of development) and the conditions within which it exists (e.g. climate) change. Something which is often easier within a SuDS than a traditional system relying on a fully enclosed pipe network.

These criteria are associated with a series of design standards that describe how the system should function (as summarised in Table 4.3), many of which have been defined within the guidance in a quantified or semi-quantified way (Woods Ballard et al. 2015).

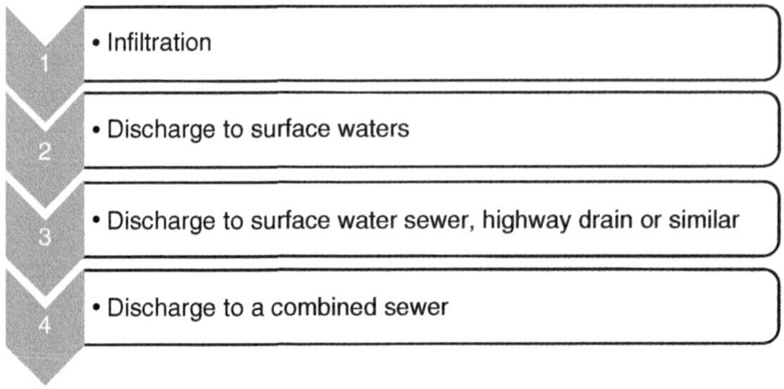

Fig. 4.4 Hierarchy of SuDS discharge options. (Based on Woods Ballard et al. 2015)

Table 4.3 Summarised design standards for water quantity management

Volume control
For frequent rainfall events – SuDS design should ensure that discharge does not occur for the majority of small rainfall events
For extreme rainfall events – SuDS design should be capable of controlling runoff from the site during extreme events (usually up to the 1 in 100 year event)
Peak control
During events likely to impact on the morphology, ecology or capacity of receiving waters (often taken as the 1 in 1 year event) or the capacity of receiving sewers – SuDS design should constrain peak flow rates to greenfield rates for the same return period
During extreme events – SuDS design should constrain peak flow rates to greenfield rates for the same event
Flood risk control
SuDS capacity design – Should be sufficient to prevent flooding (except where flooding forms part of the design) up to a predefined design event (usually at least the 1 in 30 year event)
Exceedance capacity design – For events that exceed the design capacity, the risks associated with flooding should be determined and managed accordingly

Based on Woods Ballard et al. (2015)

Water Quality Within a SuDS, natural cleaning processes (chemical, physical and biological) are controlled and managed in such a way as to ensure that receiving waters are protected in a sustainable manner. There are, however, a bewildering (and sometimes unknown) number of pollutants that could be picked up in urban environments, some of which will be deposited within the catchment on a more or less constant basis (e.g. oils and heavy metals from road traffic), whilst others may only be present in the event of an accidental spillage. In as far as pollutants that build up on urban surfaces are concerned, build-up is dependent on the rate of deposition, the rate of breakdown in the environment and the length of the antecedent dry period (i.e. the length of time since the pollutants were last washed from the surface). Mobilisation is then determined by the intensity of a rainfall event, the properties of the pollutant (e.g. whether it's soluble or particulate) and the connectivity of the drainage system (i.e. how efficiently it moves from source to receiving waters). Other pollutants, such as those from misconnections or accidental spillages, may be deposited on surfaces or directly into the sewerage system and then either move towards a receiving water immediately or sit in the system for some time for mobilisation during a rainfall event (see Sect. 4.4.1). This, however, is not the end of the story, since the risk to the environment is also a function of the sensitivity (e.g. whether the ecosystem contains sensitive species) and size (potential for dilution) of the receiving waterbody. As a result, both contaminant delivery and actual impact are highly site and pollutant dependent, with some substances resulting in short-term acute impacts (i.e. poisoning events) and others long-term chronic effects (such as those resulting from a general decline in dissolve oxygen levels due to organic matter deposition). The risks posed to groundwaters are also site specific, being dependent on the type of contaminant (persistence in the environment), rate of infiltration and the level of connectivity between the pollutant source and receiving groundwater body (e.g. whether they are separated by an impermeable layer). Should they become polluted with a persistent contaminant however, they can effectively become permanently damaged, since although methods for aquifer cleaning exist, they are very costly and not always suitable for use.

To deal with this potentially complex problem, the UK's SuDS Manual establishes two broad design criteria (Woods Ballard et al. 2015):

- Firstly, to support the management of water quality in receiving waters (surface and groundwaters), by ensuring that both runoff and infiltrated water are of a sufficiently high standard, even if the receiving waters are already of a poor standard (so as to ensure that the effectiveness of any future improvement programme is not limited by the development in question). This is achieved through some combination of the approaches shown in Fig. 4.5 (see Sect. 4.2 for example systems).
- Secondly, as in the case of water quantity (and for the same reasons), to design system resilience to cope with future change.

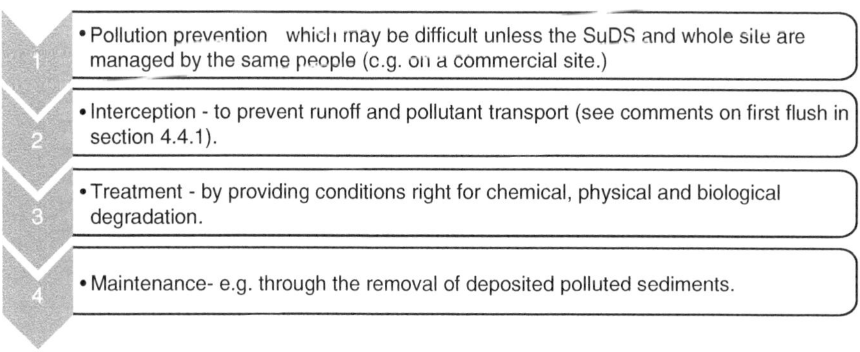

Fig. 4.5 SuDS approaches to water quality management. (Based on Woods Ballard et al. 2015)

Table 4.4 Summarised design standards for water quality management

Water quality
For frequent rainfall events
As for water quantity, a SuDS design should ensure that discharge does not occur for the majority of small rainfall events, so as to prevent pollutant transfer
For larger rainfall events
SuDS design should be capable of treating runoff sufficiently to prevent negative impacts in receiving waters (surface or groundwaters)

Based on Woods Ballard et al. (2015)

These criteria too are associated with design standards which describe how the system should function (as summarised in Table 4.4), many of which have been defined within the guidance in a quantified or semi-quantified way (Woods Ballard et al. 2015).

Amenity Although often a subsidiary goal, good urban design should contribute to the quality of life of the population living and working in or visiting the area. Water can be a valuable resource in this respect (Woods Ballard et al. 2015), as evidenced by the number of urban developments that incorporate some form of water feature. Many SuDS components have the potential to serve a number of purposes which contribute to the amenity value of an area, including through the provision of green space, pleasing environmental features, urban cooling (see Sect. 4.3) and a general feeling of wellbeing. Consequently, although guidance for the UK (Woods Ballard et al. 2015) doesn't contain any specific design standards for amenity provision (amenity value is difficult to quantify), it does contain a number of more general design criteria (Woods Ballard et al. 2015):

- Firstly, to maximise multifunctionality. Although water quantity and/or quality are likely to be the key drivers behind a SuDS, many SuDS components keep and manage water at the surface, meaning that (unlike subsurface drainage systems) they can perform a number of other roles, and this property should be taken advantage of wherever possible/practical.
- Secondly, to enhance the visual character (aesthetics) of an area.
- Thirdly, to ensure safe surface water management. SuDS are no more or less dangerous than other surface water systems; however, whenever water is kept above ground, there are potential risks to health and safety, particularly if those waters are polluted with, for example, microbiological contaminants. Consequently such issues (and in some cases public concerns) must be considered at the design stage.
- Fourthly, to support development resilience/adaptability to future change. In terms of the wider urban area (rather than the SuDS itself), SuDS can contribute to resilience/adaptability through such things as urban cooling, helping to mitigate the future impacts of climate change and water resource provision contributing to the general sustainability of water consumption.

- Fifthly, to maximise legibility. The surface nature of many SuDS components means that they can play a valuable role in informing the public about the way in which water is being managed, giving them a degree of 'ownership' of the system and an interest in ensuring that it continues to function appropriately.
- Sixthly, to support community environmental learning. Again, the fact that many SuDS components are above ground means that some at least can function as educational resources allowing the community to learn both about the functioning of catchments and wider environmental issues (e.g. aquatic biodiversity).

Biodiversity In many ways the biodiversity element of SuDS design is related to amenity, in that the provision of habitat also contributes to several amenity objectives. Nevertheless, there are a number of specific requirements for maximising biodiversity benefits, which are distinct from those of amenity. Being surface-based, many SuDS components have characteristics that are intrinsically very valuable to both aquatic and terrestrial plants and animals, and UK guidance establishes design criteria aimed at maximising that value (Woods Ballard et al. 2015):

- Firstly, to support and protect natural local habitats and species. In as far as practical, habitats developed as part of a SuDS should be as close as possible to those that would naturally be present in the area and should complement those that are already present and as a result support native fauna and flora.
- Secondly, to contribute to local biodiversity objectives. In some ways this is related to the above but recognises that in many areas, there are already habitats and species that have been identified as priorities (e.g. in local Biodiversity Action Plans, BAPs), and so SuDS should, where possible, aid in delivering on those pre-existing objectives.
- Thirdly, to contribute to habitat connectivity. Many SuDS have some degree of linear form as a result of being intended to manage water from its source (generally where it falls as rain) to receiving waters. This provides an excellent opportunity to connect habitats in such a way that their overall benefit is greater than the sum of the parts. Urbanisation has a tendency to leave isolated pockets of valuable habitat, which are often too small to maintain sustainable populations of some species on their own. When connected by linear habitats, these become more resilient (see below) as a result of allowing animals (in the main) to move over a wider area and minimising the threats posed by disease and/or environmental stresses (e.g. drought).
- Fourthly, to create diverse, self-sustaining and resilient ecosystems. Where possible, SuDS should incorporate a variety of components (as most do), resulting in a diversity of habitats that maximises biodiversity benefits. The best SuDS therefore incorporate structural and species variety in terms of vegetation and a range of water depths (including areas that are temporarily wet) and for greatest benefit should be protected from excessive pollution (something that can be difficult where the primary goal is the treatment of polluted waters). It is also important to aim for ecosystems to become as self-sustaining as possible, so as to minimise the need for ongoing intervention, and for those ecosystems to be

capable of performing their biodiversity role even in the face of future changes, such as those likely to be associated with climate change. Where management is required, this should be carried out in a manner that is sympathetic to the needs of biodiversity.

4.4.4 Field Implementation of SuDS

It is beyond the scope of this volume to cover in detail all the elements considered to be part of the SuDS design process (Woods Ballard et al. 2015; AECOM 2013, etc.); the main UK source on the matter (Woods Ballard et al. 2015) however, divides it into four stages, albeit that most are comprised of a number of sub-steps (Fig. 4.6). Other guidance documents may vary this somewhat, but nevertheless the general approach is often similar (e.g. AECOM 2013).

1. *Setting strategic surface water management objectives:* The first requirement is to define the strategic surface water management objectives of the development in question (a similar requirement exists for retrofitting projects), in relation to such things as runoff quality, flood risk, biodiversity, amenity, climate change resilience and so on. This sets the overall goals towards which a SuDS may contribute and therefore the framework within which all subsequent design steps operate.
2. *Conceptual design:* At this stage, the goal is to identify a series of SuDS components for inclusion in the management trains of the various parts of the area being developed/renewed (Woods Ballard et al. 2015). Feeding into this is the need for a clear understanding of both the catchment within which the SuDS is intended to function (topography, flow paths, soils, geology, climate, current land use, etc.) and the properties of the proposed development (building types, proposed land use, intensity of use, etc.). This provides a clear foundation on which to base a definition of the objectives of the SuDS in terms of the four pillars described above, which is related to, but not the same as, the strategic surface water management objectives, and indeed it may go further (Woods Ballard et al. 2015). This, together with a knowledge of the ultimate points of discharge (infiltration, surface waters, sewers, etc. – see above), allows management trains to be developed for each sub-catchment of the area (small developments may be comprised of a single sub-catchment), in line with the aim of treating runoff close to the point of origin and composed of SuDS components performing some or all of the water harvesting (see Chap. 6), runoff interception/infiltration, runoff storage and water transfer roles (Woods Ballard et al. 2015).
3. *Outline design:* In the UK, the production of an outline design is often an input to the process of obtaining outline planning permission, although even where this isn't done, the various steps will still be required before those in the detailed design stage can be completed (Woods Ballard et al. 2015). In essence it is the process of putting the meat on the bones of the conceptual design, by sizing the

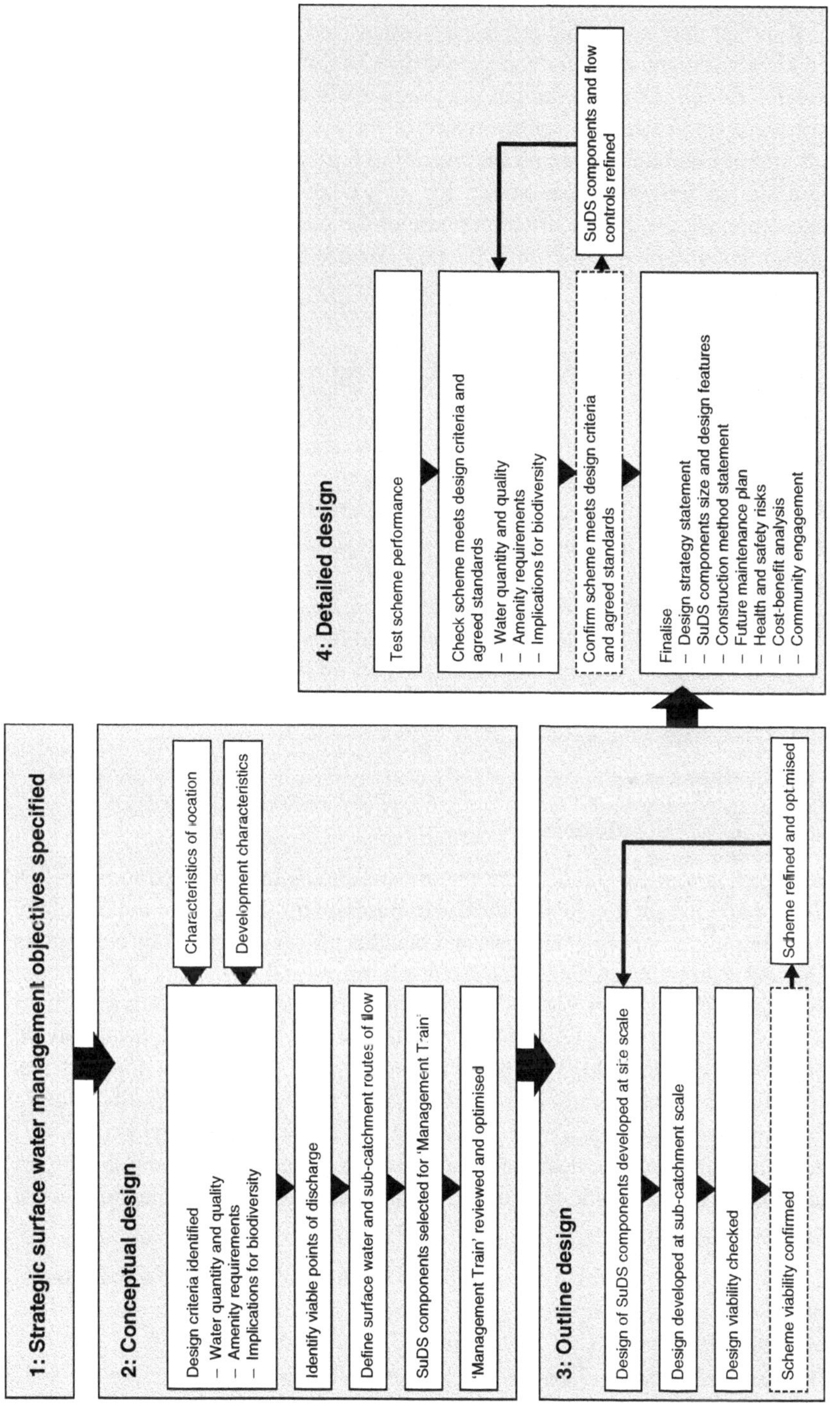

Fig. 4.6 SuDS design stages. (Based on Woods Ballard et al. 2015)

various components (based on the expected runoff and pollutant delivery rates, etc.), so that they are both capable of meeting the design criteria/specifications and have sufficient scope for accommodating future change.
4. *Detailed design:* Detailed designs are produced in order to obtain full planning permission and require the performance of the system to be tested and (just as importantly) demonstrated and the overall expected performance compared to the design criteria/specifications set for it. It is also at this stage that detailed plans are made for the construction phase of the project and any ongoing maintenance that will be required subsequently (Woods Ballard et al. 2015).

4.5 SuDS Life Cycle Assessment (LCA)

It is generally accepted that the capital costs associated with SuDS are likely to be lower than those for conventional drainage systems, as there is less need for the costly installation of subsurface sewerage pipes and related infrastructure; however, the requirement for ongoing maintenance may be significantly higher (HR Wallingford 2004). Consequently, it is recognised that if informed decisions are to be made about the merits of SuDS implementation, then the costs and benefits, particularly in comparison with the alternatives, need to be assessed and demonstrated (HR Wallingford 2004; Sharma 2008). In a report on doing so however (published in 2013), the UK's Construction Industry Research and Information Association concluded that there were as yet (CIRIA 2013):

> "...no comprehensive tools or techniques being used anywhere in the world that provide the reliability and validity needed for a robust estimate of the added benefits of SuDS, especially for a monetised assessment."

Nevertheless, it was also highlighted that existing tools may be used to provide a way forwards. One such system, Life Cycle Assessment (LCA), is a methodology for evaluating the environmental costs and benefits of a development throughout its lifetime and is an established tool for use in environmental management. As such it has been suggested as a suitable basis on which to make such evaluations in relation to SuDS, generally by using pre-existing techniques to convert all benefits/costs into monetary values (CIRIA 2013). In the UK, for example, just such an approach has been proposed in the form of whole life costing (WLC – HR Wallingford 2004), in which financial values are assigned to costs (and benefits) of various types (Fig. 4.7) and accounted for over the design life of a project. Albeit that defining the design life accurately is problematic in relation to SuDS, since many are intended to work more or less in perpetuity.

1. *Capital costs:* may be relatively straightforward to define, particularly for traditional drainage systems, where there is a good deal of experience on which to base such assessments (HR Wallingford 2004). For SuDS there may be a less extensive track record of previous work on which to base cost estimates (HR Wallingford 2004), but this is increasing, and many of the elements in the design,

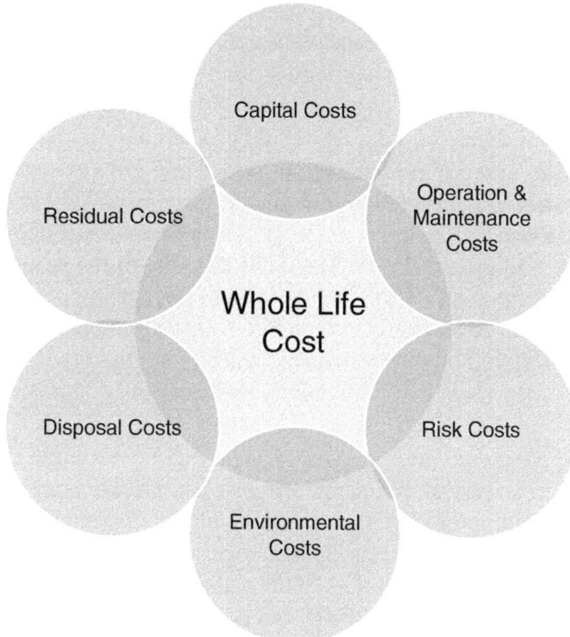

Fig. 4.7 SuDS cost elements considered in a whole life cost (WLC) assessment. (Based on HR Wallingford 2004)

planning and construction process can be costed on the basis of similar work carried out for other types of project.

2. *Operation and maintenance costs:* may be required in order to account for (HR Wallingford 2004):

   (a) *Monitoring:* is likely to be needed for both SuDS and conventional systems so as to ensure their ongoing performance (and in the case of SuDS, visual appearance), by checking for issues with the vegetation (e.g. disease/death, excessive growth), blockages, sediment build-up and so on. This is often supported by quantitative monitoring of runoff discharge and water quality, for example.
   (b) *Planned maintenance:* A degree of regular and irregular maintenance is likely to be required in the form of gully pot cleaning, vegetation cutting, sewer jetting and the removal of deposited sediments (e.g. from attenuation systems).
   (c) *Unplanned maintenance:* Although difficult to assess accurately, some maintenance will be required in order to deal with failings (or potential failings) in the system, for example, to remove more severe blockages or repair damage following extreme events, vandalism and so on.

3. *Risk costs*: are those associated with the potential damage resulting from failures in the system and are generally borne by wider society rather than the operators of a surface water drainage system (HR Wallingford 2004). For example, should a SuDS (or a conventional system) fail to prevent flooding, then the cost may be

borne by residents, insurers and the public purse. Such costs may, for example, be estimated from the expected costs of repairing flood damage or cleaning up rivers following a pollution event.

4. *Environmental (and societal) costs and benefits:* are often difficult to quantify and therefore value, not least because some are fairly subjective (e.g. the value of an amenity resource or of biodiversity). As a result the extent to which this is assessed in relation to drainage systems may be limited, although if a full economic assessment is to be carried out, then they should be included (HR Wallingford 2004). This is in part due to the fact that many of the advantages attributed to SuDS, would fall into this category, and as a result, failure to include them may result in a SuDS being significantly undervalued. They are often referred to as externalities, since they are borne by or accrue to people/groups who may have no direct involvement in the project (e.g. members of the public).
5. *Disposal costs:* may (in this context) be related to the disposal of materials that are replaced during maintenance activities (e.g. sediments, vegetation, etc. – HR Wallingford 2004), rather than the SuDS system as a whole, since as mentioned above, many are designed to operate in perpetuity.
6. *Residual costs:* refers to the value of the land used for the drainage components (HR Wallingford 2004), and as such is an assessment of the income foregone by not using the land for something else, with many SuDS systems requiring considerably more above ground space in which to operate. For example, swales require an area of land to be set aside for them in a way that subsurface sewerage generally does not.

4.6 Knowledge Gaps

It is clear from the feelings expressed in a 2016 survey of professionals and practitioners working in SuDS design, implementation and approval across the UK (Hydro International 2016) that the widespread uptake of SuDS faces a number of barriers. For example, it was clear that there was still considerable reliance on the use of proprietary SuDS components within schemes (Hydro International 2016), suggesting that many in the industry have a lack of confidence in the ability of soft options to deliver the required level of reliable flood control and, in particular, water quality management. Indeed, in the above survey, more than half of the respondents felt that greater use of proprietary systems to protect water quality would lead to greater use of SuDS schemes. In some cases such an approach may be entirely justified, but there are those who felt that proprietary systems should only be used as a last resort once other options have been exhausted (Hydro International 2016), when in fact they appear to be being used to provide confidence in the performance of systems. Many of those surveyed agreed that uncertainty around the maintenance requirements and performance of SuDS components presented a barrier to schemes being adopted (Hydro International 2016), which in turn makes it difficult for developers to incorporate SuDS in their designs. This lack of confidence may reflect a perceived lack of a suitable knowledge base of data on system performance in a range of situations, something which is gradually being addressed as ever greater numbers of systems enter service, and clearly there is a need that this should continue. In addition however, where such data is recorded, it should be made widely available, so that others can draw on that knowledge in developing their own SuDS.

4.7 Conclusions

Traditional surface water and combined sewerage systems, designed to collect and transfer precipitation away from its point of origin as quickly and efficiently as possible, have a number of disadvantages compared to SuDS, including limited ability to treat water quality and a lack of resilience to change (e.g. increased urbanisation, climate change). Consequently, SuDS have many features that recommend them to developers and local authorities alike; indeed there is now a considerable emphasis on supporting the uptake of SuDS technologies in policies and legislation across the UK. Despite this however, there are many in the industry who feel that there is a lack of commitment on the part of government to the long-term delivery of SuDS (Hydro International 2016), although in part at least this probably reflects a feeling that insufficient funds and other resources have been committed to flood resilience in the UK in general (SuDS based or otherwise – Hydro International 2016). Nevertheless, the number of component options for inclusion in SuDS management trains is substantial, such that it should be possible to identify suitable combinations for many situations, although retrofitting in existing drainage systems can still be problematic. A lack of confidence in the performance and ongoing management requirements of SuDS, however, means that many developers and those charged with approving projects are reluctant to rely solely on soft options without the inclusion of proprietary systems within the management train.

References

AECOM (2013) Water, people, places: a guide for master planning sustainable drainage into developments. AECOM, London

Allen A, Chapman D (2001) Impacts of afforestation on groundwater resources and quality. Hydrol J 9(4):390–400

Anglian Water Services (2011) Towards sustainable water stewardship: Sustainable Drainage Systems (SUDS) adoption manual. Anglian Water Services Ltd, Huntingdon

Ashley R, Blanksby J, Chapman J et al (2007) Towards integrated approaches to reduce flood risk in urban areas. In: Ashley R, Garvin S, Pasche E et al (eds) Advances in urban flood management. Taylor and Francis, London, pp 415–432

Barbosa AE, Fernandes JN, David LM (2012) Key issues for sustainable urban stormwater management. Water Res 46(20):6787–6798

Bertrand-Krajewski J, Chebbo G, Saget A (1998) Distribution of pollutant mass vs volume in stormwater discharges and the first flush phenomenon. Water Res 32(8):2341–2356
Charlesworth SM, Harker E, Rickard S (2003) A review of sustainable drainage systems (SuDS): a soft option for hard drainage questions. Geography 88(2):99–107
CIRIA (2013) Demonstrating the multiple benefits of SuDS – a business case (phase 2): draft literature review. CIRIA, London
City of Austin (1990) The first flush of runoff and its effects on control structure design. City of Austin, Austin
DCLG – Department for Communities and Local Government (2012) National planning policy framework. DCLG, London
Department of the Environment (Northern Ireland) (2014) Planning and flood risk – revised planning policy statement 15. Department of the Environment, Belfast
Department of the Environment (Northern Ireland) (2015) Strategic planning policy statement for Northern Ireland (SPPS): planning for sustainable development. Department of the Environment, Belfast
Ellis JB, Shutes RB, Revitt DM (2003) Guidance manual for constructed wetlands: RandD technical report P2-159/TR2. Environment Agency, Bristol
Essex County Council (2016) Sustainable drainage systems: design guide. Essex County Council, Chelmsford
European Council (1991) Council Directive of 21 May 1991 Concerning Urban Wastewater Treatment (91/271/EC). J Eur Commun L 135:40–52
European Parliament and the Council of the European Union (2000) Directive 2000/60/EC of the European Parliament and of the Council of 23 October 2000 establishing a framework for community action in the field of water policy. J Eur Commun L 327:1–72
European Parliament and the Council of the European Union (2007) Directive 2007/60/EC of the European Parliament and of the Council of 23 October 2007 on the assessment and management of flood risks. J Eur Commun L 288:27–34
Fletcher TD, Andrieu H, Hamel P (2013) Understanding, management and modelling of urban hydrology and its consequences for receiving waters: a state of the art. Adv Water Resour 51:261–279
Gassert F, Reig P, Luo T et al (2013) A weighted aggregation of spatially distinct hydrological indicators: aqueduct country and river basin rankings. World Resources Institute, Washington, DC
Geiger J (2002) Is all your rain going down the drain? Urban forest research July 2002. Center for Urban Forest Research, Davis
Holman-Dodds JK, Bradley AA, Potter KW (2003) Evaluation of hydrologic benefits of infiltration based urban storm water management. J Am Water Resour Assoc 39(1):205–215
HR Wallingford (2004) Whole life costing for sustainable drainage: report SR 627. HR Wallingford, Wallingford
Hydro International (2016) Engineering nature's way – SuDS: the state of the nation 2016. Hydro International, Clevedon. http://www.engineeringnaturesway.co.uk/resource/suds-the-state-of-the-nation-2016-survey-report/. Accessed 12 Jan 2018
Jenkins GJ, Murphy JM, Sexton DM et al (2009) UK climate projections: briefing report. Met Office Hadley Centre, Exeter
Lee JH, Bang KW, Ketchum LH et al (2002) First flush analysis of urban storm runoff. Sci Total Environ 293(1–3):163–175
Miller JD, Hutchins M (2017) The impacts of urbanisation and climate change on urban flooding and urban water quality: a review of the evidence concerning the United Kingdom. J Hydrol Reg Stud 12:345–362
Miller JD, Kim H, Kjeldsen TR et al (2014) Assessing the impact of urbanization on storm runoff in a peri-urban catchment using historical change in impervious cover. J Hydrol 515:59–70
Nascimento NO, Ellis JB, Baptista MB et al (1999) Using detention basins: operational experience and lessons. Urban Water 1:113–124

National Assembly for Wales (2004) Planning policy wales – technical advice note 15: development and flood risk. Welsh Assembly Government, Cardiff

Newton J, Gedge D, Early P et al (2007) Building greener: guidance on the use of green roofs, green walls and complementary features on buildings. CIRIA, London

Northern Ireland Environment Agency (2011) Managing stormwater: a strategy for promoting the use of sustainable drainage systems (SuDS) within Northern Ireland. Northern Ireland Environment Agency, Belfast

Robert Bray Associates (2012) Promoting sustainable drainage systems: design guidance for Islington. Islington Council, London

Scottish Government (2014a) Ambition, opportunity, place: Scotland's third national planning framework. The Scottish Government, Edinburgh

Scottish Government (2014b) Scottish planning policy. The Scottish Government, Edinburgh

Scottish Water (2015) Sewers for Scotland – a technical specification for the design and construction of sewerage infrastructure, 3rd edn. Scottish Water, Stepps

Sharma D (2008) Sustainable drainage system (SuDs) for stormwater management: a technological and policy intervention to combat diffuse pollution. Proceedings 11th international conference on urban drainage, Edinburgh, Scotland

Tourbier JT, White I (2007) Sustainable measures for flood attenuation: sustainable drainage and conveyance systems SUDACS. In: Ashley R, Garvin S, Pasche E et al (eds) Advances in urban flood management. Taylor and Francis, London, pp 13–28

UN DESA – Population Division (2015) World urbanization prospects: the 2014 revision, (ST/ESA/SER.A/366). UN Department of Economic and Social Affairs, New York

UN DESA – Population Division (2017) World population prospects: the 2017 revision, volume I: comprehensive tables (ST/ESA/SER.A/399). UN DESA, New York

US EPA (2005) National management measures to control nonpoint source pollution from urban areas. US EPA, Washington, DC

van Dijk AIJM, Keenan RJ (2007) Planted forests and water in perspective. For Ecol Manag 251(1–2):1–9

Welsh Government (2016) Planning policy wales: edition 9. Welsh Government, Cardiff

Wilson S, Bray B, Neesam S et al (2009) Sustainable drainage: Cambridge design and adoption guide. Cambridge City Council, Cambridge

Woods Ballard B, Wilson S, Udale-Clarke H et al (2015) The SuDS manual. CIRIA, London

Zhou Q (2014) A review of sustainable urban drainage systems considering the climate change and urbanization impacts. Water 6(4):975–992

Chapter 5
Recycling and Treatment of Water Under Urban Intensification

Rupak Aryal, Arturo Aburto Medina, Andrew S. Ball, Veeriah Jegatheesan, Felicity Roddick, Jaya Kandasamy, and Saravanamuthu Vigneswaran

Abstract With the ever increasing population growth in urban areas, stakeholders have adopted numerous water sensitive urban design (WSUD) measures to enable the recycling and reuse of stormwater for a range of non-potable fit-for-purpose uses. This chapter highlights the most common measures adopted across Australia in the recent past, their benefits and limitations. Findings suggest that the adoption of WSUD measures have provided multiple tangible and intangible benefits from a social, environmental and economic context. For further expansion and adoption of WSUD measures, the multiple benefits need to be communicated and shared with the scientific and the broader community further to create sustainable and water resilient urban areas.

Keywords Water sensitive urban design (WSUD) · Urban hydrology · Urban water demand · Water quality · Recycling

5.1 Introduction

Over the last 30 years, the population in the world has doubled, and the number of people living in urban areas has surpassed the number living in rural areas. Australia is at the forefront of this urban intensification. Rapid urban development in major cities across Australia has introduced numerous adverse changes to the natural environment, and these problems are compounding. It is estimated that more than 90%

R. Aryal · J. Kandasamy · S. Vigneswaran
School of Civil and Environmental Engineering, University of Technology, Sydney, Australia

A. Aburto Medina (✉) · A. S. Ball
Centre for Environmental Sustainability and Remediation and School of Science, RMIT University, Bundoora, VIC, Australia
e-mail: arturoaburto.medina@rmit.edu.au

V. Jegatheesan · F. Roddick
Water: Effective Technologies and Tools Research Centre, School of Engineering, RMIT University, Melbourne, VIC, Australia

V. Jegatheesan et al. (eds.), *Urban Stormwater and Flood Management*, Applied Environmental Science and Engineering for a Sustainable Future,
https://doi.org/10.1007/978-3-030-11818-1_5

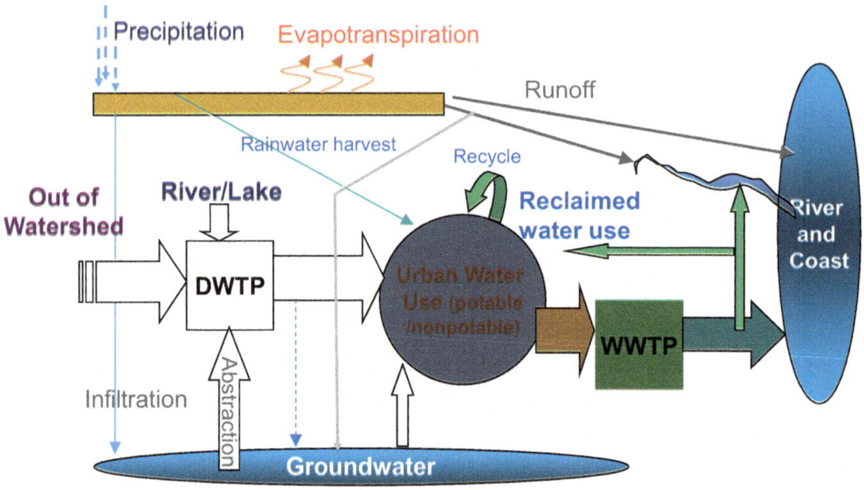

Fig. 5.1 Urban water balance. (Furumai 2008)

of the Australian population will live in cities by 2050 (ABS 2013b). This data itself explains the challenges inherent in managing the urban water demand in cities and for translating these challenges into opportunities for sustainable and smart cities. One of the major impacts of population growth is on the urban hydrological cycle. This includes fresh water management, urban peak flow management and also urban water recycling. In Australia, the average daily water use per capita is around 230 L, and wastewater discharge is around 130 L per capita per day. By 2030 it is estimated that in Sydney alone, the additional water demand will exceed 140 ML, and extra wastewater generated will exceed 65 ML (GHD 2012). Along with the population growth, there are significant changes to the landscape in urban areas and a rise in the conversion of pervious areas to impervious areas. The impervious area in a typical urban environment will often constitute more than 60% of the total area (Lu and Weng 2006). This increases the translation of rainfall to stormwater runoff and changes to the physical, chemical and biological characteristics of stormwater. Quantifying the impacts of urban development on hydrological systems is difficult compared to quantifying the physical, chemical and biological changes that stormwater runoff undergo (Gergel et al. 2002; Allan 2004; Chadwick et al. 2006).

Figure 5.1 illustrates the urban water balance that includes groundwater abstraction and recharge.

5.2 Water Recycling and Reuse in Australia

In the past, urban water management in Australia primarily focussed on measures for protection against flooding (Wong 2006). With growing urbanisation, this focus has changed. An understanding of the value of freshwater following a decade of drought (2000–2010), the pollution caused by urban stormwater runoff and its impact on human and ecological health, the focus of the last few decades has shifted towards developing and applying Water Sensitive Urban Design (WSUD) approaches for water harvesting and reuse and incorporating it into the urban fabric resulting in changes to policy, planning and design (Breen et al. 2006; Fletcher et al. 2006; Wong 2006).

In Australia, the management of stormwater is under the jurisdiction of local government. The regulation of water recycling and reuse in Australia is complex, differing from state to state and sometimes varying between local governments within the same state. Guidelines applicable to the management of urban runoff in Australia can be broadly separated into (1) runoff management through engineered water sensitive urban design and (2) protection of human health for reuse.

The Australian national stormwater harvesting guidelines were introduced in 2009 to protect human health. It originated from the second phase of the recycling guidelines, the *Australian Guidelines for Water Recycling: Managing Health and Environmental Risks (Phase 2) Stormwater Harvesting and Reuse* (AGWR-SHR) (NRMMC et al. 2009b). The Guideline is applicable to new, small to medium urban water reuse systems involving non-potable end use. The end uses include:

- Irrigation of sporting fields, golf courses, bowling greens, parks and gardens
- Dual reticulation for indoor and outdoor use (e.g. toilet flushing, laundry use, irrigation of garden food crops and ornamental gardens)
- Firefighting
- Irrigation of commercial food crops
- Irrigation of non-food crops (e.g. trees, turf, woodlots and flowers)

AGWR-SHR provide guidance for managing risks associated stormwater in relation to pathogens in the form of log reduction targets for three reference organisms including *Rotavirus*, *Cryptosporidium* and *Campylobacter jejuni*. However, the guideline does not provide guidance in relation to chemical compounds potentially present in stormwater. The most applicable Australian Guideline for assessment of chemical hazards is the approach outlined in the *Australian Guidelines for Water Recycling: Managing Health and Environmental Risks (Phase 1)* (AGWR-MHER) (NRMMC et al. 2006), whilst managing chemical compounds in urban runoff is outlined in the *Australian Guidelines for Water Recycling: Managing Health and Environmental Risks (Phase 2) – Augmentation of drinking water supplies* AGWR-ADW (NRMMC et al. 2008). Guideline threshold values provided for chemical compounds are limited to heavy metals, nutrients and a group of polycyclic aromatic hydrocarbons (NRMMC et al. 2009b).

The water quality guidelines most applicable to managed aquifer recharge (MAR) schemes are determined by the source water and the end use application (NRMMC et al. 2009a). Potential source waters for aquifer recharge include stormwater, water recycled from wastewater treatment plants, water from streams and lakes, groundwater drawn from other aquifers or drawn remotely from the same aquifer and water from drinking water distribution systems, including desalinated seawater (NRMMC et al. 2009a). End use applications include drinking water supply, irrigation, industrial purposes and environmental purposes such as increasing base flows and maintaining lake ecosystem functions (NRMMC et al. 2009a).

5.3 Water Sensitive Urban Design (WSUD)

Across Australia and internationally, a growing body of urban water professionals are focused on transitioning from conventional stormwater drainage systems to more sustainable urban water management (SUWM) in order to respond to the challenges posed the rapidly growing urban population and the impacts of climate change and the consequent environmental degradation. In the last three decades, a range of WSUD approaches have been adopted in both small precincts and at larger scale to managing urban stormwater runoff and protecting the receiving water environment from pollution, flooding, erosion and scouring, by reducing peak flow rate and total runoff volume, as well as the provision of an alternative non-potable water supply source (McAlister 1999; DEC 2006). WSUD has been recognised as a tool to minimise the disruption to catchment hydrology as a consequence of urbanisation.

The increasing implementation of WSUD measures is strengthened by an increased awareness of climate change and its consequences among professionals and the community. It confirms that further progress can be made if both technical and philosophical approaches are clearly understood. Many advocates have raised concerns from time to time about the slow adoption of WSUD in some regions, particularly related to innovation and incorporation of more sustainable technologies, not clearly understanding that policies and administrative hurdles are always challenging during implementation (Brown 2008). There is no clear regulatory guidance especially for stormwater, relating to chemical compounds and pathogens that are potentially present or remain in the water prior to reuse. Consequently, the human and ecosystem health risks posed by chemical compounds and pathogens, which are usually assessed through guideline indicators that may not necessarily be applicable to a broad range of hazards present in the urban stormwater available for recycling and reuse, could result in lack of public confidence. Reliable treatment technologies, storage and supply options and life cycle costing methodologies are the other hurdles. Furthermore, the adoption of new technologies requires additional investment, which is critical in the integration of recycled water as part of urban water supply.

The overall uptake of WSUD across Australia has added value to water recycling and reuse to make cities water resilient and also to tackle increasing urban water pollution. Although WSUD measures are adopted to minimise environmental problems with end use purpose, there is still inadequate monitoring and evaluation of the performance of these systems. Table 5.1 provides examples of WSUD policy approaches adopted in Australia. The table reflects the differences in the implementation of WSUD objectives across the country.

Table 5.1 WSUD policy features across Australia[a]

Policy components	NSW	VIC	QLD	WA	SA	ACT	NT	TAS
WSUD targets	√	√	√	√	√	√	√	√
Integrated planning requirements		√	√	√		√		
Local WSUD guidelines	√	√	√	√	√		√	√
Legislated WSUD requirement for development	√ (selected areas)	√	√	√		√		√
Central coordination at state level		√	√	√		√		√
Capacity building programme	√	√	√	√	(In progress)	√	√ (limited progress)	√

[a]Note: adopted from *WSUD Impediments and Potential: Contributions to the SA Urban Water blueprint (Goyder Institute for Water Research Technical Report Series No. 14/16)*

Some of the most common WSUD applications adopted in Australia are discussed below.

5.4 Rainwater Harvesting (RWH)

In Australia, RWH has been an important source of potable water since the 1800s in rural/farm areas or areas where there is no reticulated water supply. The basic design of household rainwater systems includes collection of water from roofs via gravity into storage tanks, where it can then be drawn to meet household water demand.

Over the last few decades, interest in RWH has increased due to water stress in many cities (Imteaz et al. 2011; Imteaz et al. 2014; Haque et al. 2016; OEH 2018). Rainwater harvesting has been made mandatory for new dwellings (e.g. Building Sustainability Index, BASIX policy in New South Wales) in many major cities in Australia. Through these, and especially restrictions to reticulated water supply to households during the millennium drought, the adoption of RWH increased dramatically. In Australia, 34% of households have adopted RWH systems as of 2017, which increased from 24% in 2007 to 32% in 2010 (Christian Amos et al. 2016). Around 47% of Adelaide city households have a rainwater tank installed at their dwelling, followed by Brisbane (44%) and Melbourne (31%). Among households, over half (51%) have the tank connected to a tap or outlet inside the dwelling (ABS 2013a).

RWH can contribute to decreasing stormwater runoff peak and volume. In a study by Tsai and Chiu (2012) in Taipei City, reductions of 26.5–100% in runoff volume and 15–100% in peak flow were noted. Zhang et al. (2012) observed reductions of 14–58% in runoff volume and to help mitigate urban water-logging problems in a residential district in Nanjing, China. A similar outcome has been reported by Campisano et al. (2014) in Catania, Italy.

There are a number of advantages and limitations in rainwater harvesting.

Advantages

(i) The utilisation of rainwater collection and reuse systems for non-potable purposes can significantly reduce potable water consumption.
(ii) The collection of rainwater runoff reduces site discharge and pollutant loads receiving waters.
(iii) The collection and reuse of rainwater runoff attenuate flood peaks.

Limitations

(i) Roof collection systems tend to be less effective for water supply in areas that have hot and dry climatic conditions for prolonged periods.
(ii) The water quality needs to be monitored regularly.
(iii) The time period for recovery of costs on RWH systems could be relatively long.

5.5 Managed Aquifer Recharge

Managed aquifer recharge (MAR) is the planned infiltration or injection of water into an aquifer during times when water is available and the subsequent recovery of the water from the same well when needed (Pyne 1995; Martin and Dillion 2002; Sheng 2005). The recharge water may be treated wastewater or stormwater. Before recharge, the stormwater needs to be treated to prevent the aquifer from becoming clogged with particulate or organic material or contaminated by pollutants. MAR offers the opportunity to use water resources more efficiently and to prolong the available supply of water. MAR increases the yield of the aquifer and protects it from seawater intrusion. Where suitable aquifers are available, MAR offers a potentially low-cost method of storing water underground as an alternative to surface reservoirs. MAR is also useful for aquifers that have experienced decline in the water table because of high extraction and can help to restore groundwater levels if sufficient water is recharged.

There are a number of methods to recharge groundwater through MAR. These include injection and recovery through a single well (both recharge and recovery through the same well) or multiple wells (to attenuate contaminants), through infiltration basins where water is collected in constructed ponds and allowed to infiltrate through the base of the pond to an underlying shallow aquifer and through bank filtration where pumping wells adjacent to a water course are used to draw water from a stream into the aquifer. MAR has gained acceptance as a water resource management technique in several parts of Australia (Fig. 5.2).

Some advantages and limitations of ASR are listed below.

Advantages

(i) Aquifer storage is the most feasible approach to recharge groundwater after abstraction.

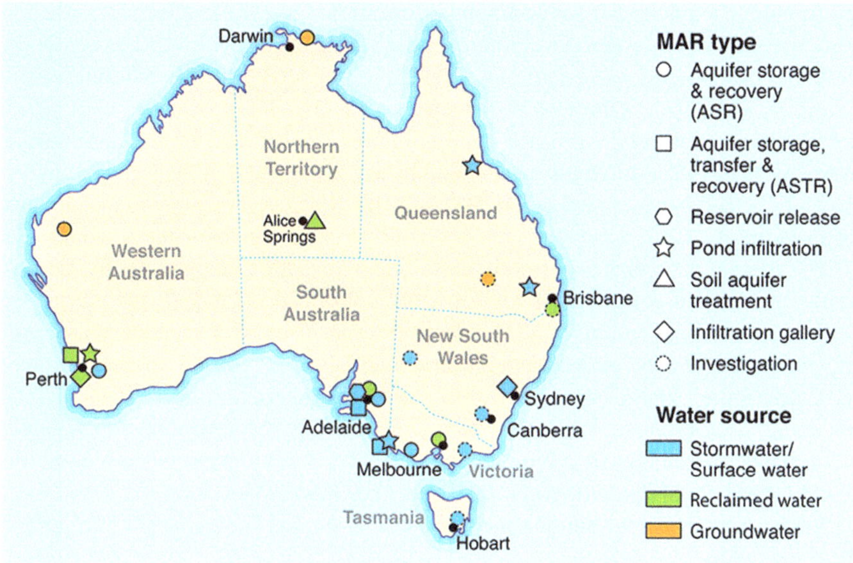

Fig. 5.2 A snapshot of MAR types in Australia in 2015 (CSIRO https://research.csiro.au/mar/using-managed-aquifer-recharge/#map)

(ii) Aquifer storage can be a reliable source of water for drought-prone areas where rainfall is relatively limited.

Limitations

(i) Water destined for aquifer storage needs to be treated beforehand.
(ii) Injection of water may add pollutants to the groundwater.
(iii) Clogging of the well may occur if water is not properly treated prior to injection.

5.6 Irrigation

Treated wastewater (recycled water) has been not widely utilised in Australia, although its use began in the mid-1990s (Anderson et al. 2006; Seshadri et al. 2015) in many states such as New South Wales (NSW), Queensland and Victoria, whereas some states have taken the lead to maximise recycled water use such as South Australia and Western Australia. Primary reasons for this underutilisation are stringent regulations relating to the use of recycled water and lack of public acceptance (Dolnicar et al. 2014; Hurlimann and Dolnicar 2016). Regulations governing the use of reclaimed water are not uniform throughout Australia. Each state and territory has guidelines and regulations for managing natural resources and public health in their jurisdiction. Legislation for wastewater reuse is covered by acts relating to food safety, public health and/or environmental protection. Over the last decade, there has been a steady increase in the volume of reused water in agriculture, and currently around 11.5% of total wastewater generated is reused (ABS 2013a). Although agriculture uses the largest amount of recycled water (103 GL/year), use of recycled water accounts for only 1% of the total volume of water used by the agricultural sector (BOM 2015). Among the states, South Australia (SA) is at the forefront in the use of reclaimed water for irrigation. Recently, the SA government announced plans for an additional 12 GL/year to irrigate horticulture in Northern Adelaide, an increase of 60% compared to what is currently used. Use in irrigation includes for food crops (grown for human consumption and possibly consumed uncooked, such as green leafy vegetables and fruits), non-food crops or crops used after secondary processing (fodder, fibre, seed crops, pastures, commercial nurseries, sod farms, commercial aquaculture). One of the major obstacles in using recycled water is its quality and being fit for purpose. The main water quality factors that determine the suitability of recycled water for irrigation are the presence of pathogens, salinity, sodicity, toxicity, trace metals and nutrients (Lazarova and Bahri 2004).

Some advantages and limitations of the use of treated wastewater for irrigation are listed below.

Advantages

(i) A regular source of water for irrigation.
(ii) Reuse water can be more economical.
(iii) Reuse water can be a good source of nutrients for agriculture.

Limitations

(i) Health and safety can be a concern.
(ii) Chlorine and salt content can have an adverse impact on soil structure.
(iii) Pathogens and other pollutants can be a concern in relation to human health.
(iv) High turbidity can cause blockage and slime formation, especially in drip irrigation systems.

5.7 Case Studies

The case studies presented below provide examples of adoption of water recycling and reuse.

A. Rainwater Harvesting *Figtree Place*, in Hamilton, inner suburban Newcastle, has adopted integrated stormwater management in a residential and commercial setting. The site, consisting of 27 residential units, employs rainwater tanks, infiltration trenches and a central basin in which treated stormwater enters an unconfined aquifer. Reinforced concrete underground rainwater tanks are used. The ‘first flush’ pit associated with four rainwater tanks is designed to separate the first 2 mm of rainfall from the inflow. Around 60% total water saving could be achieved through complete stormwater runoff retention. No exceedances of guideline values for metals and chemical parameters have been noted in the tank water. However, sampling of roof water using 50 L drum and automatic sampler revealed exceedances for metals, other chemical compounds and bacteria. Concentrations of bacteria in roof water entering rainwater tanks were typically two orders of magnitude greater than concentrations found in tank water.

NSW Building Sustainability Index (BASIX) Due to the growing population in Sydney and across NSW, the 2010 Metropolitan Water Plan for the greater Sydney region (Sydney, Illawara and the Blue Mountain area) was introduced. This has an emphasis on water security via recycling along with reservoirs, desalination and water efficiency. The plan placed alternative water use as a high priority for investment and adoption by the community which resulted in water conservation programmes for new dwellings being embedded in planning through BASIX. Recently, the Metropolitan Water Plan 2017 was released which emphasises the facilitation of a more integrated approach to providing water, wastewater and stormwater services, which will contribute to making communities more liveable and resilient and moving towards becoming WaterSmart cities. The BASIX programme has helped to increase the collection of rainwater and its use for various domestic purposes including toilet flushing and garden irrigation and, consequently, reduce potable water consumption in individual households by 29% by 2015. This achievement was lower than the initial expectation of 40% reduction, and it was realised that adoption of BASIX was slower than expected (SydneyWater 2017).

B. Salisbury Constructed Wetlands The Salisbury constructed wetlands (Fig. 5.3) are a good example of urban water reuse. The City of Salisbury has more than 50 wetlands to maximise multiple benefits that include:

- Restoring habitat and increasing biodiversity
- Providing flood protection
- Providing natural filtration, cleansing of stormwater, and enabling a low-cost treatment option for reuse
- Protecting the sensitive downstream Barker Inlet, an estuary into the Gulf of St Vincent and the largest fish breeding nursery in South Australia
- Creating attractive landscape features
- Providing areas for passive recreation and enjoyment
- Enabling research and development
- Providing opportunities for environmental education and awareness
- Significantly contributing to the ultimate goal of a sustainable urban environment

C. Alexandra and Queen Victoria Gardens' Ponds and Lakes The Alexandra and Queen Victoria Gardens provide almost 10 ha of green space in the heart of Melbourne City. The gardens have a stormwater harvesting system that captures, treats and stores 20 ML of stormwater each year for irrigation and ornamental use.

Fig. 5.3 Wetland in Salisbury North Adelaide

The innovative system makes use of retrofitted ornamental ponds in the Queen Victoria Gardens for storing stormwater. Overall, the system reduces potable water use in Alexandra Gardens and Queen Victoria Gardens by 55%.

Water is diverted from the existing drains beneath Queen Victoria Gardens into a gross pollutant trap where gross pollutants such as leaves and litter are removed. Next, a long sedimentation chamber settles out small particles, including fine sand and oil. The water is then transferred to a series of three ponds for storage. When the water level in the ponds reaches a predefined threshold, the water is pumped under Alexandra Avenue to Alexandra Gardens for treatment where a biofilter is employed to remove pollutants as the water seeps through the soil and plant root systems. The treated water is collected under the bed in a pump well and transferred to an above ground storage tank. The tank stores 230 KL of water for irrigation. Excess clean water is returned to the Yarra River via the stormwater drains.

D. Sydney Olympic Park The Water Reclamation and Management Scheme at Sydney Olympic Park represents a large-scale approach to recycling non-potable water. Established in 2000, the aim is to provide all the water required for toilet flushing, irrigation and other residential uses in the park and the nearby suburb of Newington. The scheme conserves approximately 850 ML of mains water per year. Stormwater is captured in the Brickpit Reservoir, having a 300 ML storage capacity, and a series of freshwater wetlands constructed as part of the Haslams Creek area remediation. Treatment through the wetlands reduces sediment and nutrient loads up to 90%. Stormwater from the storages is combined with reclaimed water 'mined' from a trunk sewer, filtered via continuous microfiltration and disinfected prior to use. A dual reticulation system distributes the water to the park and to Newington homes. In addition to conserving water, implementation of the scheme has allowed for the annual diversion of approximately 550 ML of sewage normally discharged through ocean outfalls.

E. Managed Aquifer Recharge (MAR) The Northern Adelaide MAR (Parafield stormwater harvesting) scheme in the Salisbury area, South Australia, is part of an integrated network for managing stormwater. The scheme was developed for harvesting urban stormwater in the City of Salisbury and utilises wetland treatment followed by aquifer storage and recovery. The water is collected from a number of catchments via a main pipe and transferred to the harvesting site.

The maximum rate of injection is 8 ML/day (Dandy et al. 2013) with potential to increase to 60 GL/y by 2050 (Dandy et al. 2013; Mankad et al. 2013). The main consumers of the Parafield scheme are households at Mawson Lakes, who use the water for garden watering and toilet flushing, and G.H. Michell and Sons, who use the water as part of the wool scouring process.

Table 5.2 summarises various WSUD adopted in Australia, their success and impediments.

Table 5.2 Summary of adoption of WSUD

System	What makes them successful	Main limitations	Capital cost, maintenance costs, energy costs	Skills required to run such schemes	Adoption hurdle with respect to recycled water (if any)
RWH	Raw rainwater relatively clean and suitable for non-potable purposes. Easily implemented at household scale. Has long history of use in Australia	Treatment required for potable use.	Low	Low	Suitable treatment required for potable use.
Constructed wetlands and ponds	Uses natural treatment processes. Creates an aesthetic environment. Easily adopted into new land releases	Requires relatively large land area. Treatment required for potable uses	High capital cost, moderate low maintenance cost and energy costs	Moderate	Suitable treatment required for potable use
Landscape and gardening irrigation	Provides continuous water source for agriculture in dry climates. Water quality standards for use are available and not onerous to adopt	Pollutants present can be a potential human and ecosystem health hazard	Relatively high infrastructure costs	Moderate	Suitable treatment required for potable use

(continued)

Table 5.2 (continued)

System	What makes them successful	Main limitations	Capital cost, maintenance costs, energy costs	Skills required to run such schemes	Adoption hurdle with respect to recycled water (if any)
Managed aquifer recharge (MAR)	Availability of large volumes of water for a range of uses	Treatment required before aquifer recharge	High capital cost, maintenance cost and energy costs	High	Appropriate hydrogeological conditions needed

5.8 Conclusions

The uptake water recycling across Australia has been at various scales, at allotment, precinct and large industrial scale. Precinct and large-scale adoption has been made possible due to local councils, state governments or national-scale initiation and funding. At the allotment scale, systems have been adopted by developers and individual households to be a part of local and state-wide initiatives or to meet local or state government policies and sometimes motivated by an individual's environmental awareness.

The concept of water sensitive urban design (WSUD) plays a key enabling role in water recycling. Application of WSUD is often multi-objective as it attempts to address stormwater management, flood mitigation, urban water reuse, reduction in pollutant loads to receiving water bodies, safeguarding the ecosystem and improving landscape amenity.

However, there are a number of constraints that need to be addressed for better adoption of stormwater reuse. These include:

- Improved capacity for reuse and WSUD adoption and implementation
  The uptake of WSUD in the early days was primarily to manage water flows and to improve water quality close to the predevelopment hydrology. This has evolved into the management of urban water under the conditions of changing climate and urban water stress. Planning, design and implementation has progressed markedly. However, there are still insufficient studies and field data generated in relation to the treatment performance of existing WSUD schemes. This has created a lack of understanding of the importance of WSUD under urban intensification and has also delayed the enhancement of WSUD approaches.
  Education is required for the community to understand the importance of WSUD and the ability to enhance urban liveability. For example, adoption of rainwater harvesting has been widespread due to regulatory requirements. However, the community may not fully understand how the use of harvested rainwater within the household can positively influence not only the demand for reticulated water but also the downstream stream ecology. Making the community aware of the key benefits can build community acceptance and support to ensure widespread adoption.

- Integration and coordination across management boundaries
  Water recycling extends over a broad range of discipline that encompasses policy to professional and management to community issues. Therefore, there is a strong need to consider how water recycling is integrated across different sectors in a consistent way that helps to achieve multiple objectives such as flood risk management, safeguarding and human and ecosystem health, as an alternative to mains water supply and enhance landscape amenity.
- More research to bridge the knowledge gaps in relation to performance, operation and management of WSUD systems
  There is limited knowledge in relation to the performance, socio-economic benefits, life cycle costs and operation and management of WSUD systems. This acts as a major impediment to its wide acceptance and adoption.
- Identification of the risks and costs associated with reuse
  The adoption of water reuse at any scale requires a shift in the traditional approach to managing urban water. The conventional approach to water supply has been practised for a long time and is well understood, and the life cycle cost can be reasonably calculated. In the case of reuse, many stakeholders are reluctant to adopt such systems due to numerous knowledge gaps. These include its potential multiple tangible and intangible benefits, associated risk and cost and uncertainty in the policy and regulatory environment.

References

ABS (2013a) Environmental issues: water use and conservation, Mar 2013. Australian Bureau of Statistics

ABS (2013b) Population projections Australia. In: Australian Bureau of Statistics (ed) Australian Bureau of Statistics, Canberra

Allan JD (2004) Landscapes and riverscapes: the influence of land use on stream ecosystems. Annu Rev Ecol Evol SystAnnu Rev Ecol Evol Syst 35:257–284

Anderson J, Davis C, Stevens D, Kelly J, McLaughlin M, Unkovich M (2006) Reclaimed water use in Australia: an overview of Australia and reclaimed water. In: Stevens D (ed) Growing crops with reclaimed wastewater. CSIRO Publishing, Collingwood 5

BOM (2015) Water in Australia 2014–2015. http://www.bom.gov.au/water/waterinaustralia/files/Water-in-Australia-2013-14.pdf. Bureau of Meteorology Australia, Melbourne, VIC 30001, p 68

Breen P, Wong THF, Lawrence I (2006) Constructed wetlands and ponds. In: Wong THF (ed) Australian runoff quality: a guide to water sensitive urban design. Engineers Media, Crows Nest

Brown RR (2008) Local institutional development and organizational change for advancing sustainable urban water futures. Environ Manag 41:221–233

Campisano A, Di Liberto D, Modica C, Reitano S (2014) Potential for peak flow reduction by rainwater harvesting tanks. Procedia Eng 89:1507–1514

Chadwick MA, Dobberfuhl DR, Benke AC, Huryn AD, Suberkropp K, Thiele JE (2006) Urbanization affects stream ecosystem function by altering hydrology, chemistry, and biotic richness. Ecol Appl 16:1796–1807

Christian Amos C, Rahman A, Mwangi Gathenya J (2016) Economic analysis and feasibility of rainwater harvesting systems in urban and peri-urban environments: a review of the global situation with a special focus on Australia and Kenya. Water 8:149

Dandy G, Ganji A, Kandulu J, Hatton MacDonald D, Marchi A, Maier H, Mankad A, Schmidt C (2013) Managed aquifer recharge and stormwater use options: net benefits report. Goyder Institute for Water Research (ed). Adelaide

DEC (2006) Managing urban stormwater: harvesting and reuse. Department of Environment and Conservation (New South Wales)

Dolnicar S, Hurlimann A, Gruen B (2014) Corrigendum to "What affects public acceptance of recycled and desalinated water?" [Water Res. 45 (2) (2011) pp. 933–943]. Water Res 64:321

Fletcher TD, Wong THF, Breen P (2006) Buffer strips, vegetated swales and bioretention systems. In: Wong THF (ed) Australian runoff quality: a guide to water sensitive urban design. Engineers Media, Crows Nest

Furumai H (2008) Rainwater and reclaimed wastewater for sustainable urban water use. Phys Chem Earth, Parts A/B/C 33:340–346

Gergel SE, Turner MG, Miller JR, Melack JM, Stanley EH (2002) Landscape indicators of human impacts to riverine systems. Aquat Sci 64:118–128

GHD (2012) Recycled water plan. https://www.cityofsydney.nsw.gov.au/__data/assets/pdf_file/0007/151198/Recycled-Water-Plan.pdf. City of Sydney, Sydney

Haque MM, Rahman A, Samali B (2016) Evaluation of climate change impacts on rainwater harvesting. J Clean Prod 137:60–69

Hurlimann A, Dolnicar S (2016) Public acceptance and perceptions of alternative water sources: a comparative study in nine locations. Int J Water Resour Dev 32:650–673

Imteaz MA, Ahsan A, Naser J, Rahman A (2011) Reliability analysis of rainwater tanks in Melbourne using daily water balance model. Resour Conserv Recycl 56:80–86

Imteaz MA, Matos C, Shanableh A (2014) Impacts of climatic variability on rainwater tank outcomes for an inland city, Canberra. Int J Hydrol Sci Technol 4:177–191

Lazarova V, Bahri A (2004) Water reuse for irrigation: agriculture, landscapes, and turf grass. CRC Press, Boca Raton

Lu D, Weng Q (2006) Use of impervious surface in urban land-use classification. Remote Sens Environ 102:146–160

Mankad A, Walton A, Leonard R (2013) Public attitudes towards managed aquifer recharge and stormwater use in Adelaide. Goyder Institute for Water Research Technical Report Series, Adelaide

Martin R, Dillion P (2002) Aquifer storage and recovery. Department of Water, Land and Biodiversity Conservation

McAlister T (1999) Stormwater reuse: a balanced assessment. Stormwater Industry Association Conference, Sydney

NRMMC, EPHC, NHMRC (2006) Australian guidelines for water recycling: managing health and environmental risks (Phase 1)

NRMMC, EPHC, NHMRC (2008) Australian guidelines for water recycling: managing health and environmental risks (Phase 2) Augmentation of drinking water supplies. Natural Resource Management Ministerial Council, the Environment Protection and Health Council and the National Health and Medical Research Council

NRMMC, EPHC, NHMRC (2009a) Australian guidelines for water recycling: managing health and environmental risks (Phase 2) Managed aquifer recharge. Natural Resource Management Ministerial Council, the Environment Protection and Health Council and the National Health and Medical Research Council

NRMMC, EPHC, NHMRC (2009b) Australian guidelines for water recycling: managing health and environmental risks (Phase 2) Stormwater harvesting and reuse. Natural Resource Management Ministerial Council, the Environment Protection and Health Council and the National Health and Medical Research Council

OEH (2018) Office of environment and heritage-home saver rebates program data. https://www.environment.nsw.gov.au/energyefficiencyindustry/home-data-rebates.htm
Pyne RDG (1995) Groundwater recharge and wells: a guide to aquifer storage recovery. CRC press, Boca Raton
Seshadri B, Bolan NS, Kunhikrishnan A, Chowdhury S, Thangarajan R, Chuasavathi T (2015) Recycled water irrigation in Australia. In: Thangavel P, Sridevi G (eds) Environmental sustainability: role of green technologies. Springer India, New Delhi, pp 39–48
Sheng Z (2005) An aquifer storage and recovery system with reclaimed wastewater to preserve native groundwater resources in El Paso, Texas. J Environ Manag 75:367–377
SydneyWater (2017) Water conservation strategy. Sydney
Tsai Y, Chiu Y (2012) Urban storm runoff utilization strategy and volume analysis in multipurpose rainwater detention-retention systems. Adv Water Sci 23:341–345
Wong T (2006) Australian runoff quality: a guide to water sensitive urban design
Zhang X, Hu M, Chen G, Xu Y (2012) Urban rainwater utilization and its role in mitigating urban waterlogging problems—a case study in Nanjing, China. Water Resour Manag 26:3757–3766

Chapter 6
Stormwater Harvesting

Kathy Lewis, Doug Warner, and John Tzilivakis

Abstract Stormwater harvesting is not as widely adopted in the UK as it is in Australia or some other parts of Europe such as Germany, despite the potential for such systems. Where it is used, professional standard systems are more likely to be installed on privately owned, eco-designed newly built constructions as opposed to standard residential properties. There is some evidence that UK government policies are strengthening in this area but appear to be driven through sustainable drainage strategies for flood mitigation, rather than for supplementing a traditional centrally supplied potable water system. The obstacles to increased uptake are largely cost related, although issues relating to appropriate enforceable regulations, public perception of risk and lack of technical awareness are also evident. Whilst the benefits for water security and the environment are not questionable, the financial viability in the UK is still in doubt due to the high cost of installing these systems, especially when retrofitted. Sound policies, initiatives and incentives are required to increase uptake in line with other countries at the forefront of these technologies.

Keywords Rainwater harvesting · Retrofitting · Roof surface material · Soakaway · Storage tank · Stormwater harvesting · Surface run-off · Water governance · Water storage · Water treatment

6.1 Introduction

Whilst the UK is not as water stressed as some other countries, its rapidly growing population and associated urban developments are putting severe pressure on water resources. According to the Office for National Statistics (ONS), the UK's population grew to an estimated 65.1 million in 2015, which represents an increase of just over half a million people against the 2014 figure (ONS 2017). Each year in the UK,

K. Lewis · D. Warner (✉) · J. Tzilivakis
Agriculture and Environment Research Unit (AERU), University of Hertfordshire, Hatfield, Hertfordshire, UK
e-mail: d.j.warner@herts.ac.uk

V. Jegatheesan et al. (eds.), *Urban Stormwater and Flood Management*, Applied Environmental Science and Engineering for a Sustainable Future,
https://doi.org/10.1007/978-3-030-11818-1_6

around 18 billion tonnes of water are extracted from rivers, groundwater aquifers and reservoirs for use by domestic households and industry and for electricity generation and agricultural production. Water consumption in the UK has grown by an average of 1% annually since 1930 (Waterwise 2012), with domestic households being by far the greatest consumer, accounting for 6 billion tonnes annually (Waterwise 2012). Around 30% of this is used for toilet flushing and the transport of foul waste to the sewer (Fewkes 1999). There is also little doubt that the UK is becoming more urban, and it is well known that as urban areas expand, water demand rises (Domene and Sauri 2006; Wijitkosum and Sriburi 2008). Evidence of this can be found from CORINE 2012 data, which shows that over 2250 km^2, equivalent to 1% of the UK's total area, underwent land use change between 2006 and 2012. Much of this was due to the conversion of forest to artificial surfaces (buildings, roads, car parking surfaces, etc.), indicating urban expansion (Geographical 2017). It is anticipated that continued growth in population, climate change and food security concerns will increase demand for water still further in the coming years. In addition, considerable amounts of taxpayer money are spent on the provision of public water for building, maintaining, operating and replacing the supporting infrastructures (Arpke and Hutzler 2006; USDE 2006). Consequently, there is a growing need for the UK to find ways of adapting to and mitigating water insecurity.

Stormwater harvesting or 'short-cutting the hydrological cycle' (Way et al. 2010) to reduce storm run-off and provide harvested rainwater for domestic, non-potable uses is nothing new, having been around in one form or another for thousands of years (Gould and Nissen-Petersen 1999; Campisano et al. 2017). As such, it offers one option for mitigating water insecurity, potentially reducing reliance on conventional supplies, but as yet this is not a technology that has been adopted to any significant extent in the UK. UK policy on water harvesting is still maturing, and in some respects there is a lack of clarity, although it is clear that the government recognises that water collection will benefit the environment, particularly in terms of flood management and biodiversity protection, as well as supplementing current traditional water supplies (Brown 2005; Wheater and Evans 2009; Ward 2010; London Assembly 2016). Hence over the last decade or so, sustainable water strategies have begun to emerge. For example, the UK's Department for Environment, Food and Rural Affairs (Defra) 'future water' strategy aims to reduce domestic water consumption to 130 L $person^{-1}$ day^{-1} by the year 2030 (Defra 2008), and the UK Sustainable Buildings Task Group (2004) prioritised water conservation as one of the four key areas to improve the sustainability of new and refurbished buildings. In recent years, stormwater harvesting and run-off water management have also featured more prominently as part of sustainable drainage policies (Spillett et al. 2005; Woods-Ballard et al. 2007) and regulations, for example, the UK's Flood and Water Management Act, 2010. The government has also offered various incentives for water harvesting projects to industry. For example, the agricultural industry has been encouraged to build reservoirs, install underground storage tanks and take other water conservation measures. Grants for these types of projects have been available through the Countryside Stewardship Scheme. In addition, the Enhanced

Capital Allowance scheme allows businesses to claim back the tax associated with the purchase of certain water harvesting equipment against taxable profits in the year of purchase.

An interesting observation is that in some countries and scientific fora, there are two distinct terms that refer to the harvesting of water: 'stormwater harvesting' and 'rainwater harvesting'. The former term is used to refer specifically to the collection of ground surface run-off water, and so the collected water has the potential to be significantly contaminated with particulates, oils, pesticides and nitrates to name but a few likely pollutants, whereas the latter term refers specifically to the collection of rainwater, predominately from roofs, and, whilst still polluted, is likely to be cleaner than stormwater. In the UK, the definition of the two terms is quite blurred, as rainwater harvesting has also been used to refer to the collection of water from, for example, a purpose-built hard standing that may have residential or public uses such as a patio or vehicle parking area (Environment Agency 2010).

Despite the general concepts being well established and the undeniable drivers for its adoption, the uptake of stormwater harvesting technologies in the UK and the techniques' move from novel to mainstream have been hampered by a number of technical, financial and social barriers (Ward 2010; Melville-Shreeve et al. 2016a). Indeed, UK installations are few and far between, and, where such systems have been installed, these have largely been rainwater collection systems, particularly from privately owned and iconic public buildings, rather than that from stormwater (ground surface run-off water) specifically. Berndtsson (2004) expressed concern that the UK had fallen behind other countries in the uptake of domestic rainwater harvesting on a large scale. In 2005 Hassell (2005) stated that the market for rainwater harvesting in the UK was increasing rapidly. However, over a decade later, its uptake in the UK has not, according to some authors, progressed significantly, being limited to an estimated 7500–9000 households (Fewkes 2012; Ward et al. 2014), a relatively minor contribution considering the number of potential households within the UK. The potential for rainwater harvesting for non-potable uses in the UK is noted by several authors (e.g. Wheater and Evans 2009; Environment Agency 2010; Fewkes 2012; Melville-Shreeve et al. 2016a, b, c), with benefits to be realised, particularly in the more water-scarce south-east of the UK, which go beyond that of water conservation and environmental protection. For example, the chalk aquifers from which much of the mains supplied water is extracted within the south-east catchment cause it to be 'hard' (high in calcium carbonate ($CaCO_3^-$). Hard water is responsible for damage to pipes and domestic appliances (washing machines, boilers, etc.) with a further associated economic cost to consumers. The use of harvested rainwater for these non-potable applications removes such issues.

6.2 Stormwater Harvesting Approaches in the UK

The uptake of rainwater harvesting in the UK is not limited by a lack of available options (Heggen 2000; Leggett and Shaffer 2002). In the past, both Brown (2005) and the Environment Agency (2010) singled out the approaches favourable in

Germany and Australia for implementing successful strategies, citing them as models to which the UK should aspire. In Germany, more than 35% of new buildings have rainwater collection systems installed. Frankfurt airport is a good example, where rainwater from the roof of the new terminal building is collected and subsequently used for toilet flushing, watering plants and garden areas and other non-potable uses. Water is collected in 100 m^3 capacity tanks located in the airport basement (Trautner 2001). Similarly the Technical University of Darmstadt has a large rainwater harvesting system, which saves around 80,000 m^3 of water each year and which is used for toilet flushing and supplied to laboratories for cooling and cleaning purposes.

In practice, often due to space considerations for storing the collected water, the approach used tends to depend on if the system is a new build or if it is to be retrofitted. At a basic level, all rainwater harvesting systems operate using a similar concept. Rainwater flows off the roof into the guttering and down pipes before passing through a filter to remove leaves and other debris. The rainwater is then collected in a storage tank, which may or may not include a pump and filter. Fewkes (2012), summarising an article by Herrmann and Schmida (1999), outlined four main variations to this basic system that have potential for use in the UK:

(a) *Total flow*: All of the run-off enters the storage tank through a filter mechanism; excess water overflows from the storage tank into the local drainage system as shown in Fig. 6.1a. These types of system are often small scale and so can be retrofitted to, for example, garages, sheds and conservatories. The water can then be used directly for garden irrigation, car washing, etc.
(b) *Diverter system*: This is similar to the total flow system, but only a proportion of the total run-off is diverted into a storage tank; the remainder bypasses it, going to the drainage system as shown in Fig. 6.1b. The quantity of water diverted depends on the run-off flow rate.
(c) *Retention and throttle type*: This approach requires additional storage (retention volume), which retains run-off during periods of high flow, and then is emptied during low flow periods via a throttle valve as shown in Fig. 6.1c.
(d) *Infiltration type*: A storage tank overflow infiltrates adjacent ground, which acts as a 'soakaway', allowing water table recharge as shown in Fig. 6.1d.

With respect to residential, public and commercial buildings, systems to conserve water supplies are more commonly being considered by architects and builders at the design and build. Regardless of if it is a new build or a retrofit, there are various regulations that must be complied with. These include controls embedded within the UK's Building Regulations and Water Supply (Water Fittings) Regulations, 1999, predominately to ensure that these, potentially polluted, water resources do not become mixed with purified domestic water supplies and so endanger public health (Table 6.1). There are also a series of British Standards related to compliance with regulations and the adoption of best practice (Table 6.1).

With respect to residential properties, at the very basic level, most houses with garden space will use a water butt to collect roof run-off water for use in the garden. Indeed, several of the UK's water companies provide these at a discounted rate to

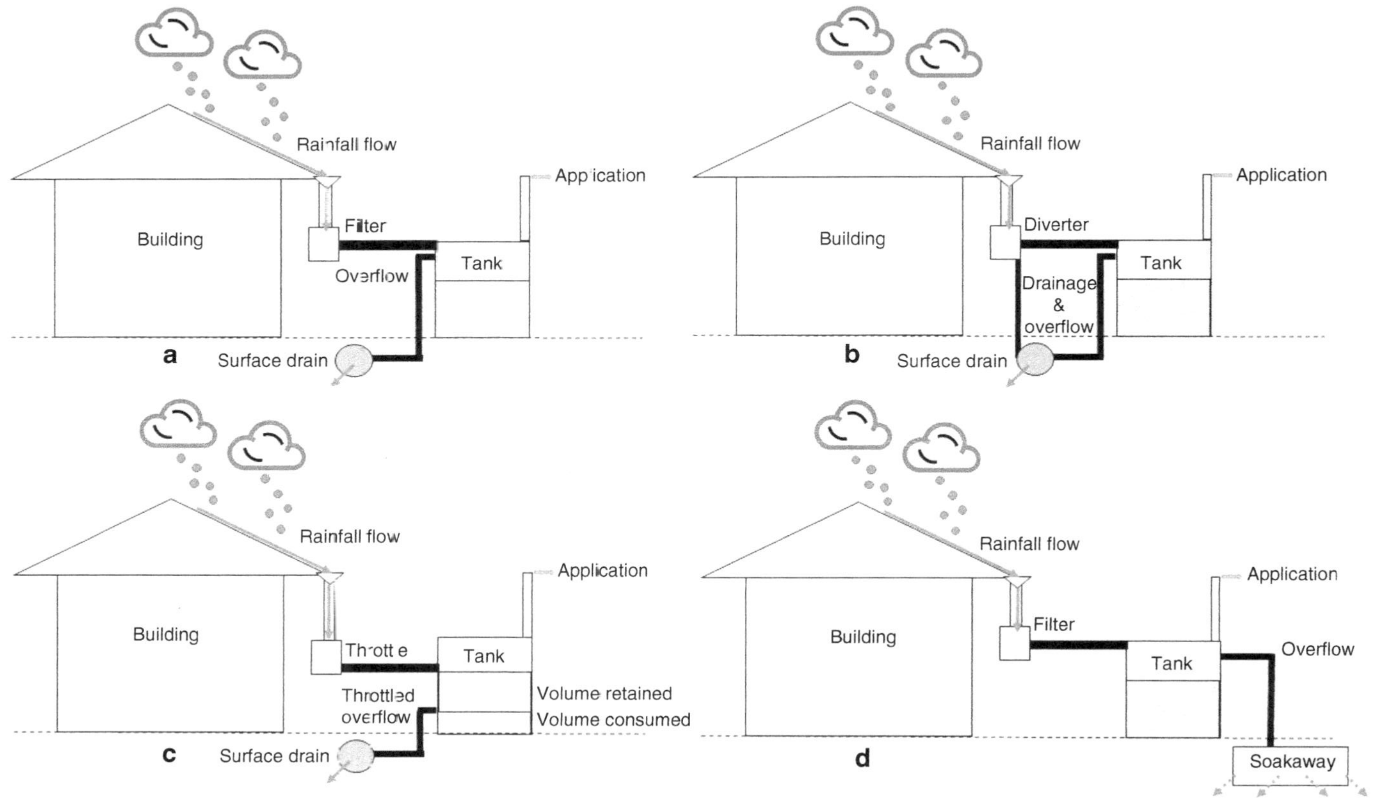

Fig. 6.1 Types of rainwater harvesting systems with potential for use in the UK: (**a**) total flow arrangement, (**b**) diverter arrangement, (**c**) retention and throttle system, (**d**) infiltration type. (Based on Nouh and Al-Shamsy 2001)

Table 6.1 Regulations and standards relevant to rainwater harvesting in the UK

Instrument type	Reference	Description
UK Regulation	Building Regulations 2010, as amended	Certain aspects of the UK's Building Regulations apply to the harvesting of rainwater. These include (1) the 2015 edition of Part H of the regulations on Drainage and Waste Disposal that contains information relevant to the installation of a rainwater harvesting system into a building and (2) the 2016 edition of Part G of the regulations that relates to legal requirements regarding sanitation, hot water safety and water efficiency. It also states that rainwater cannot be used as a drinking water supply
UK Regulation	Water Supply (Water Fittings) Regulations, 1999 as amended. SI 1999/1506	These regulations seek to prevent the contamination and waste of water supplied by UK water companies. Water is categorised according to quality, and regulations governing water fittings, systems and installation are category specific. SI 1999/1506 applies to England and Wales; however there are similar regulations in place in Scotland and Northern Ireland
British Standard – Code of Practice	BS 8595:2013	*Code of Practice for the Selection of Water Reuse Systems*. This document provides recommendations on how to select water reuse systems considering a variety of different factors including volumes, management, water supply and sewage infrastructure. It covers rainwater and stormwater harvesting as well as grey water reuse
British Standard – Code of Practice	BS 8515:2009+A1:2013	*Rainwater Harvesting Code of Practice*. This standard presents a code of practice that aims to protect public health by ensuring consistency of quality, installation, testing and maintenance of rainwater harvesting systems for non-potable water applications in the UK
British Standard – Code of Practice	BS 8525-1:2010	*Grey Water Systems Code of Practice*. This standard relates to the on-site collection and use of grey water as an alternative to public mains or private potable water supply. It seeks to introduce standardisation by providing recommendations on the design, installation, alteration, testing and maintenance of grey water systems utilising bathroom grey water to supply non-potable water in the UK
British Standard	BS 8525-2:2011	*System requirements for domestic grey water treatment equipment*. This standard specifies particular requirements and describes testing methods for packaged and/or site-assembled domestic grey water treatment equipment
British Standard	BS 1710:2014	*System specification for the identification of pipelines and services*. This standard specifies the colours and other information that should be used to identify pipes, ducts and electrical conduits. It requires grey water sources of water to be clearly identified via labelling

householders. However, capacity tends to be very small (>150 L) although larger ones are available (>1000 L). In terms of more professional systems, there are very few examples where these have been installed into domestic residences. One example is a housing development at Great Bow Yard, Langport in Somerset where the architects incorporated a rainwater harvesting system into an award-winning eco-design (Rainwater Harvesting Systems 2016). More usually, implementation tends to be restricted to buildings belonging to property companies and housing associations. For example, the Two Rivers Housing Association in Cinderford has installed a system that serves ten properties, having a collection tank capacity of 13,000 L (Freerain 2017). There is, however, a greater abundance of examples of larger-scale professional systems on privately owned and, to a lesser extent, public buildings, particularly those that might be classed as iconic. Table 6.2 provides a number of examples at this scale, which are currently operational in the UK.

In terms of stormwater harvesting specifically, in the UK these installations are mainly associated with sustainable drainage systems (SuDS). These systems can be used, for example, for the irrigation of urban gardens and parklands, cleaning of public vehicles and buildings and even road cleansing, thereby reducing water costs as well as reducing the risk and cost of flooding. Approaches adopted might include green roofs, vegetated open spaces, porous paving and rain gardens, for example, coupled with a water collection and delivery system. The general idea is to keep rainwater out of the local sewer system and so reduce potential sewer overflows and catchment discharges whilst retaining the water for later use (Ellis 2013). Whilst SuDS are often a part of local authority plans (e.g. Greater London Authority 2015; Hackney Council 2007), there are very few significant examples in practice.

Table 6.2 Examples of large-scale rainwater harvesting systems in operation in the UK

Type and reference	Description
Rainwater harvesting (Envirowise 2008; Environment Agency and Waterwise 2012)	At *Coolings Nurseries* near Sevenoaks in Kent, a rainwater harvesting system is reported to collect 7000 cubic metres of rainwater each year. The nursery now supplies almost 75% of its own water needs, saving over £2000/year. Run-off on to neighbouring properties has also reduced, and as a consequence flooding events locally have been much less frequent
Rainwater harvesting	There is also evidence of an increasing number of commercial plant nurseries and garden centres collecting roof water from glasshouses and similar structures for use on nursery stock, bedding and pot plants. For example, *Lowaters Nursery* uses water collected from a rainwater harvesting system in conjunction with a reservoir to irrigate plants
Rainwater harvesting (Rainwater Harvesting Systems 2016)	*Cardiff Bus depot* captures rainwater from pre-existing buildings and surrounding large areas of paved surface into an above-ground storage tank fitted with pumps, UV disinfectant systems and various management control features. The harvested water is used to flush toilets and wash vehicles. In the 6 months following installation, the system collected 447,000 L of water

(continued)

Table 6.2 (continued)

Type and reference	Description
Rainwater harvesting (Rainwater Harvesting Systems 2016)	*Gloucestershire Fire and Rescue Service* captures rainwater from various buildings, and the water is subsequently used for staff and public toilets and for washing fire response vehicles. The system includes an 18,000 L tank and is estimated to collect around 313,000 L of water annually
Rainwater harvesting	A rainwater harvesting system has been installed at the *Royal Society for the Protection of Birds education centre* at Rainham Marshes in Essex. The system uses a 6000 L storage tank and is estimated to collect around 125,000 L of water each year, which is used in the centre's toilets
Rainwater harvesting (Hurlestones 2017)	*The Shard*, a relatively new yet iconic building located near London Bridge, was built with a rainwater harvesting system. The system includes seven storage tanks, and the collected water is subsequently reused for a variety of non-potable uses
Rainwater harvesting (Hurlestones 2017)	*The Westfield Shopping Centre* in London has a rainwater harvesting system that is estimated to capture 11,000 m^3 of water annually, which is used to supply approximately 200 public toilets in the building
Waste water recycling (House of Lords 2006)	At the industrial scale, one example of wastewater reuse is that at Flag Fen gas-fired power station in Cambridgeshire. The power station is supplied by Anglian Water with high-quality treated wastewater for flue gas injection and boiler feed make-up. This saves 1.2 megalitres of potable mains water each day
Rainwater harvesting	Rainwater harvesting was a key part of the Olympic Delivery Authority's objectives to achieve a 40% reduction in potable use. Half of the site's 13,000 m^3 *Velodrome*'s roof is used to collect water, which is subsequently used for toilet flushing and landscape irrigation
Rainwater harvesting	*Several UK universities* have installed rainwater harvesting facilities on their new buildings. The University of Hertfordshire has installed them on both its Student Union and its Law buildings. The Gateway building at the University of Bedford and the Postgraduate and CPD Centre and the new library at the University's Luton campus have green sedum roofs to harvest rainwater and increase the biodiversity of the surrounding areas. Another example is the John Galsworthy Building at Kingston University that can store 10,000 l of water for use predominately in toilets
Rainwater harvesting	*London's Millennium Dome* uses a water collection system to catch the run-off from its curved structure. Water from a surface area of 90,000 m^2 is collected using an expansive guttering system, fed into a series of hoppers and finally collected in a large storage tank. The collected water is used for rest room hand basins and staff showers
Rainwater harvesting	Large capacity rainwater harvesting is undertaken at the *Museum of London*. Water from the buildings' roof is collected into 25,000 L storage tank and subsequently used in the toilets, visitor areas and to irrigate the building's green roof.

6.3 Treatment of Harvested Stormwater

Rainwater, whilst somewhat cleaner than stormwater, will still contain various pollutants, which are likely to include traces of airborne chemicals, oils and dust particulate matter as well as some animal and bird faecal material. Under the UK's Water Supply (Water Fittings) Regulations, 1999, as amended, all alternative water supplies, including collected rainwater, are classified as a fluid category 5 risk (Table 6.3) and not considered suitable for human consumption.

Risk assessments focus on identifying potential situations where the water could either contaminate the potable supply or be consumed in error. These risks are largely mitigated by the UK's Building Regulations 2010, as amended. For non-potable use, most rainwater systems do not include any purification treatment (such as biofilters) other than the inclusion of a sieve/filter to remove large particles such as leaves, other plant material and wind-blown debris. However, especially at the larger collection scale, it is not unusual for above-ground storage tanks to include UV filters to stop the formation of algae. Whilst UV filters can also improve water sanitation by removing bacteria, the consumption of harvested rainwater water is not approved in the UK, and so there is no need for harvested water to be purified in this manner other than to add a higher level of risk management. Indeed, there is little evidence that either stormwater or rainwater is treated in UK systems, independent of the normal water treatment systems operated by the regional water companies, such that it can subsequently be used for potable purposes.

Table 6.3 UK water categories governing water use applications and installations

Category 1	Pure water supplied in accordance with Section 67 of the Water industry Act 1991/ the Water Supply (Water Quality) (Scotland) Regulations 1990 and any amendment. Suitable for drinking
Category 2	Water quality is reduced due to the potential for contamination with organisms and/or materials that may alter its taste, odour or appearance, including water in hot water distribution systems. The water does not pose a threat to human health
Category 3	The water may include low concentrations of substances that may pose a threat to human health such as ethylene glycol, copper or chlorine salts
Category 4	The water may include higher concentrations of substances that may pose a significant threat to human health including chemical, carcinogenic substances or pesticides
Category 5	The water may include higher concentrations of substances that may pose a significant threat to human health, including faecal material, human or animal waste and/or pathogens from any source

6.4 Obstacles to Wider Adoption

Water harvesting in the UK has been hampered by a number of technical, financial and social barriers (Ward 2010; Melville-Shreeve et al. 2016a). Whilst the introduction of the BS 8595:2013 Code of Practice for the Selection of Water Reuse Systems (2013) is particularly important when it comes to the choice of rainwater harvesting systems, it can sometimes be an obstacle to innovation rather than an aid. For example, it stipulates that only drainage directly from roof surfaces should be collected because water originating from surface drains (i.e. stormwater) has the potential for contamination with materials such as oil (Wheater and Evans 2009). It also states that in order to prevent algal growth in stored water, storage tanks ideally should be located underground. Underground storage is also a useful approach to avoid the water freezing during the winter months (Melville-Shreeve et al. 2016d). The desire for underground storage can often limit installation in existing residential buildings due to building design and/or a lack of space. It is also an issue for the implementation of more innovative solutions such as installation below driveways and gardens. For example, a driveway or patio area made from a permeable material such as gravel, permeable asphalt or permeable block paving with a water collection system below it may not only reduce demands for potable water but also slow down the rate at which run-off water enters the drainage system, so reducing the risk of flooding. However, the code does not have statutory status.

6.4.1 Environmental and Economic Costs

Water treatment in the UK encompasses multiple techniques, each removing a particular contaminant(s). The process typically includes the passage of water through granular carbon filters, the addition of several chemical compounds (e.g. chlorine, hexafluorosilicic acid, manganese dioxide) and aeration (DWI 2009, 2016). Both the aeration process and recharge of carbon filters are energy-intensive procedures. Carbon emissions associated with mains treated water tend to fluctuate annually. A carbon equivalent (CO_2eq) of between 0.25 and 0.29 t CO_2 eq per mega litre (or kg CO_2eq L^{-1}) is cited by the Environment Agency (2008). According to lifecycle assessment (cradle to grave product evaluations), the embodied energy consumed and carbon equivalent (CO_2 eq) increased in non-potable water sourced from rainwater harvesting systems in the UK, compared to mains water (Environment Agency 2010; Parkes et al. 2010; Ward et al. 2011). The main contribution arises from the manufacture of the storage tank material, coupled with the electricity required to power the pumping system. This unfortunately contradicts the 80% target reduction in UK greenhouse gas emissions set by the UK Climate Change Act (2008). The novel rainwater harvesting systems evaluated by Melville-Shreeve et al. (2016c) that utilise low-energy pumping systems eliminate the emissions associated with the latter, making rainwater harvesting a more attractive proposition both environmentally and economically. This appears to be a critical intervention in facilitating more wide-scale uptake of rainwater harvesting in the UK and Europe as a whole.

The capacity of the storage tank is the delimiter controlling the volume and potential quantity of rainwater prevented from entering the drainage systems during extreme weather events. This, therefore, affects its flood risk mitigation potential and the quantity of mains sourced water substituted. The BS 8515:2009+A1:2013 British Standard for Rain Harvesters in the UK (2013) requires that the capacity of the rainwater harvesting system should be a minimum of either 5% of annual rainfall yield (ARY) or 5% of the annual water demand (AWD) (consumption), whichever is the smaller quantity. The volume has a significant impact on capital costs and, therefore, the viability and likelihood of installation and overall uptake within the UK. The optimal sizing of the rainwater storage tank for the supply requirements is emphasised by Fewkes (2012) and Ward et al. (2012a, b) in order to reduce cost. Overall design can also dramatically affect the financial implications of installation (Melville-Shreeve et al. 2016a). Roebuck et al. (2011) reported a rigorous financial analysis of domestic rainwater collection systems, taking a whole-life perspective. The authors concluded that harvesting rainwater was significantly less cost-effective than relying solely on mains-only water. Their data showed that domestic rainwater harvesting systems generally resulted in financial losses roughly equivalent to the installation's capital cost. Consequently, without significant financial support, domestic systems were unlikely to be cost-effective or attractive in the UK in the foreseeable future.

Melville-Shreeve et al. (2016b, c) promote the concept of 'dual-purpose' rainwater harvesting systems that capture both water for non-potable use and contain a 'sacrificial' capacity for stormwater control. Kellagher and Maneiro Franco (2007) document a 50% reduction in run-off volume for storage tanks of 1.5 m^3 capacity per person. Similar reductions are noted by Memon et al. (2009) for both combined and separate sewers. Further evidence of benefit, albeit circumstantial, is provided by Leggett et al. (2001) and Woods-Ballard et al. (2007). The main function of the system will depend on the location within the UK, with flood mitigation a greater priority in northern and western areas of the UK compared to the south-east, where water shortage mitigation increases in priority. Fewkes and Warm (2000) in testing a model of water-saving efficiency in 11 locations within the UK found it a good predictor of system performance to maximise the cost/benefit ratio. Of interest in the model is the account of annual rainfall, highly pertinent to the UK, where annual rainfall varies extensively. A storage tank could, therefore, be sized to optimise storage efficiency depending on location whilst reducing cost and payback period (Ward et al. 2012a, b), further incentivising the uptake of systems (Ward et al. 2013). It would support and facilitate the recommendation by Lash et al. (2014) that storage capacity be adapted in tandem with climate change adaptation.

The lack of capacity within the UK to retrofit existing buildings with rainwater harvesting systems compared to Australia and Germany is noted by Melville-Shreeve et al. (2016a). Retrofitting into existing buildings is quite problematic, and there is, currently little, if any, financial support or incentives of any sort to encourage uptake by the domestic sector, despite the sector's considerable water consumption levels (Melville-Shreeve et al. 2016d). These systems need space, which is often not available and can require some quite significant and thus disruptive installation activities both inside and outside of the building. Given the quantity of housing stock built before 2007 and subject to the Code for Sustainable Homes (DCLG 2008, 2010), this is a critical observation and one that needs to be addressed in the UK as a matter of priority. The cost to adapt current residential dwellings to rainwater harvesting sourced water under a 'full house system' (e.g. for lavatory flushing and use in washing machines) is also deemed prohibitive. This has been disputed to a degree in a recent study by Melville-Shreeve et al. (2016c). The inclusion of innovative rainwater harvesting systems utilising low-energy pumping systems and gravity, in addition to systems specified by the relevant British Standards published in 2013 (BS 5895:2013 and BS 8515:209+A1:2013), renders retrofitted rainwater harvesting systems more economically and environmentally viable. Grants to support the installation of rainwater harvesting systems are currently available for commercial properties under the Enhanced Capital Allowance (ECA) scheme for water, Water Technology Product List (Defra 2016). Financial investment in the water-saving technology is offset by tax relief against taxable profits for the duration of the period when the equipment is purchased (Defra 2016).

The installation of rainwater harvesting systems into new builds during the building process is not, however, currently inhibited by cost (Fewkes 2012). Whilst some building companies do install rainwater harvesting systems to their new residential builds, this is not, by any stretch of the imagination, the norm. Whilst these systems may offer social benefits associated with water conservation, installation does incur a significant financial cost that must be passed on to the buyer, be that individuals, housing associations or another body. House prices in the UK are very high and have risen sharply in recent years, mainly due to demand outstripping supply. In addition, household incomes have dropped, making houses far less affordable than they were a decade ago. Builders are being pressurised by the UK government to produce more 'affordable homes', and as a consequence, water harvesting systems are seen as a luxury and not routinely installed in order to keep developer costs low. Some authors have also identified the fact that building sustainable homes is not a priority to many building companies (SPONGE 2005). Parsons et al. (2010) reported a case where a UK Local Authority established a grant scheme to enable local developers to install water efficient technologies into new houses such that it would be no added cost to the company. Only one developer applied for the grant. Parsons et al. (2010) also concluded from their survey results that UK house builders had a poor level of technical knowledge relating to rainwater harvesting systems although there was evidence that this is improving. A 2006 House of Lords Technology Committee report (House of Lords 2006) also considered obstacles to uptake of stormwater harvesting systems in the UK, and one of those identified was the lack of strict and well-defined standards, particularly with respect to legally enforceable, sub-potable water quality standards. Some building developers have reported that they would be more comfortable installing some systems if they could prove, via compliance with standards, that the systems they had installed were fit for purpose (Parsons et al.

2010). Williams and Dair (2006) concluded that wider implementation of rainwater harvesting systems in new buildings was hampered by the construction company's concerns about being accountable if there were any issues with the system. This included greater inconvenience, increased time to complete construction or potential reputational harm. Despite this, Water Wise (2010) noted that in excess of 10,000 new domestic dwellings were constructed with water efficiency measures (albeit not specifically RWH) incorporated as standard in the East of England region between 2006 and 2009.

In theory, any future increase in cost for mains supplied potable water by WSPs (Pericli and Jenkins 2015) serves to make rainwater harvesting more attractive economically, corresponding to the tariff restructure considered necessary by Brown et al. (2010). Indeed Campisano et al. (2017) note an increase in the uptake of rainwater harvesting in several European countries, for example, Austria, Belgium, Denmark and Switzerland for this very reason. Ward et al. (2014) also highlight an issue relating to the domestic water charging system in the UK. Whilst in the past water services (e.g. water supply, infrastructure, drainage and waste water management) were charged according to the property's value, established by government to determine the level of local taxes that should be paid, an increasing number of customers now pay a consumption volumetric charge instead, as this is seen as a much fairer approach. Should rainwater harvesting become more popular, water companies would receive less income, as householders would be taking a smaller volume of water from the public water supply. However, water companies would not see a corresponding decrease in the amount of effluent they need to treat as the harvested rainwater, once used, would still be sent to the sewer in the same way as potable water. Whilst this could have serious implications for water companies and require water charging systems to be adapted, there are additional benefits to be realised that might offset this, for example, less raw water may need to be abstracted and treated. Jenkins et al. (2016) found that an increase in mains supply costs alone was insufficient to reduce domestic water consumption, albeit applicable to a small case study area in the east of England. Chappells and Medd (2008) did note, however, that consumers in the south-east of England decreased consumption in response to advice from WSPs during a period of prolonged drought in 2006. It concurs with the assertion of Brown et al. (2010) that behavioural change in the UK is unlikely unless imposed upon consumers by extreme events. An increase in drought frequency, coupled with decreased running costs, both potentially imminent, should favour an increase in the installation of rainwater harvesting systems.

6.4.2 Contamination

Other factors that potentially hinder rainwater harvesting are the contamination of the water collected. Most rainwater harvesting is derived from roof areas (Fewkes 2012), since contamination within surface drains is greater (Wheater and Evans 2009) rendering it unsuitable even for non-potable domestic use (BS 8595:2013,

2013). Optimal roof surfaces consist of inert material, such as slate, which constitute the majority of residential buildings in the UK (Fewkes 2012). Roofs that consist of bitumen, a surface typically present on older industrial units within the UK, may discolour the water which, whilst acceptable for non-potable uses, may act as a barrier with regard to social acceptance. Bitumen roof surfaces are gradually being replaced by metal roofed structures. Metal roofs tend not to discolour harvested rainwater, but the dissolution of metal ions, such as zinc, into solution may result due to the pH of rainwater being below 7 (Fewkes 2012). With the exception of water applied to gardens, harvested rainwater for use in most domestic applications would require treatment with either a disinfectant (e.g. chlorination to 0.4–0.5 mg L^{-1} free chlorine), pasteurisation via ultraviolet radiation from the sun or slow sand filtration (Helmreich and Horn 2009; Li et al. 2010). The use of sunlight is an option with greater potential in the south-east of the UK, also the area of greatest risk to water shortages.

There have also been studies on public attitudes to rainwater use. These have shown that, even for non-potable applications, there is a reluctance by many individuals to use collected water due to health concerns (e.g. Ward et al. 2008). Indeed untreated, collected rainwater will contain chemical pollutants and microbial contaminants although the exact composition and so water quality will vary considerably depending on factors such as locality and climate as well as wildlife populations that might have access to the water collection area (Fewtrell and Kay 2007; Gwenzi et al. 2015; Campisano et al. 2017). Whilst, as previously mentioned, collected rainwater can only be used for non-potable applications in the UK, there is still an associated risk, however small. There is also evidence that there is a negative public perception regarding the associated running and maintenance costs and these may also be an obstacle to uptake (Ward et al. 2010; Campisano et al. 2017). Many authors recognise the need to better incentivise domestic householders by use of schemes similar to those for renewable energy (Ward et al. 2014; Egyir et al. 2016; Campisano et al. 2017).

The flood mitigation properties of rainwater harvesting increase priority both economically and socially as justification for wide-scale adoption. Due to issues over cost and lack of capacity, Fewkes (2012) recommends prioritising the targeted retrofitting of buildings only in areas vulnerable to severe flooding. An increase in water storage volumes by a factor of 1.5–2.5 above those stipulated by the Environment Agency (2010) is also recommended to enhance their flood mitigation potential. This, however, has implications for cost (Fewkes 2012). Under the UK Flood and Water Management Act (2010), the automatic connection of drainage pipes from roofs to storm sewers is no longer standard practice in new developments. The introduction of national standards for the design, building, maintenance and management of SuDS is also provided for by the Act. Fewkes (2012) considers that the Act will promote new approaches to stormwater management and flood prevention in the UK and an area where the provision of expertise and experience from strategies currently employed in Australia will potentially be invaluable. Another important concept, absent in the UK and dealt with in detail by Hunt and

Rogers (2014), is the labelling and benchmarking of water consumption, discussed further in Chap. 8.

6.5 Knowledge Gaps

There is no paucity of experience and technical 'know-how' on the design, installation and operation of stormwater harvesting systems available, especially from countries where such systems are quite common place. This information is highly transferrable and applicable from one geographical location to another. However, for the UK specifically, there is a need for detailed cost-benefit analysis that considers multiple and combined perspectives in terms of not just water security but also food security, flood mitigation, environmental quality and biodiversity. This type of study should also consider if the wider societal benefits from water harvesting are sufficient to justify taxpayer funded capital grant schemes for both domestic home owners and commercial/industrial property owners as incentives to increase uptake. A similar study, examining the UK costs and benefits of installing large catchment rainwater harvesting systems in the more water-scarce south-east of the UK, is also needed.

6.6 Conclusion

It is evident that stormwater harvesting is not as widely adopted in the UK as it is in Australia or, indeed, in some other parts of Europe such as Germany. Where it is used, professional standard systems are more likely to be installed on privately owned eco-designed new builds than on standard residential properties. There is some evidence that government policies are strengthening in this area, but these do seem to be driven through sustainable drainage strategies for flood mitigation, rather than for supplementing traditional potable water supplies.

The obstacles to uptake are largely cost related although there are also issues relating to appropriate enforceable regulations, public perception of risk and lack of technical awareness. Whilst the benefits for water security and the environment are not questionable, the financial viability in the UK is still in doubt due to the high cost of installing these systems, especially when retrofitted. However, there is growing pressure on UK water resources, and so, sound policies, initiatives and incentives are required to increase uptake in line with other countries at the forefront of these technologies.

References

Arpke A, Hutzler N (2006) Domestic water use in the United States: a life-cycle approach? J Ind Ecol 10(1–2):169–184

Berndtsson JC (2004) Beneficial use of stormwater: a review of possibilities. Urban Water 2004(6) Chalmers University of Technology

Brown RR (2005) Impediments to integrated urban stormwater management: the need for institutional reform. Environ Manag 36(3):455–468

Brown LE, Mitchell G, Holden J et al (2010) Priority water research questions as determined by UK practitioners and policy makers. Sci Total Environ 409:256–266

Campisano A, Butler D, Ward S, Burns MJ, Friedler E, DeBusk K, Fisher-Jeffes LN, Ghisi E, Rahman A, Furumai H, Han M (2017) Urban rainwater harvesting systems: Research, implementation and future perspectives. Water Res 115:195–209

Chappells H, Medd W (2008) What is fair? Tensions between sustainable and equitable domestic water consumption in England and Wales. Local Environ 13(8):725–742

Department for Communities and Local Government (DCLG) (2008) Code for sustainable homes: technical guide. Crown Copyright UK

Department for Communities and Local Government (DCLG) (2010) Code for sustainable homes: technical guide. November 2010. Crown Copyright UK

Department for Environment, Food and Rural Affairs (Defra) (2008) Future Water – the UK's sustainable water strategy. London, UK

Department for Environment, Food and Rural Affairs (Defra) (2016) Enhanced capital allowance scheme for water Water Technology Product List. Crown copyright UK

Domene E, Saurí D (2006) Urbanisation and water consumption: influencing factors in the metropolitan region of Barcelona. Urban Stud 43(9):1605–1623

Drinking Water Inspectorate (DWI) (2009) Drinking water safety. Guidance to health and water professionals. Crown Copyright UK

Drinking Water Inspectorate (DWI) (2016) List of approved products for use in Public Water Supply in the United Kingdom. April 2016

Egyir SN, Brown C, Scott A (2016) Rainwater as a domestic water supplement in Scotland: attitudes and perceptions. Br J Environ Clim Change 6(3):160–169

Ellis JB (2013) Sustainable surface water management and green infrastructure in UK urban catchment planning. J Environ Plan Manag 56(1):24–41

Environment Agency (2008) Greenhouse gas emissions of water supply and demand management options. SC070010/SR. Environment Agency, Bristol

Environment Agency (2010) Harvesting rainwater for domestic uses: an information guide. Environment Agency, Bristol

Envirowise (2008) Reducing mains water use through rainwater harvesting: collecting a valuable free resource and reducing the impact of runoff. Envirowise Leaflet EN896, Harwell, Didcot, Oxon. http://www.wrap.org.uk/sites/files/wrap/EN896_Final.pdf. Accessed 10 Oct 2017

Fewkes A (1999) The use of rainwater for WC flushing: the field testing of a collection system. Build Environ 34:765–772

Fewkes A (2012) A review of rainwater harvesting in the UK. Struct Surv 30(2):174–194

Fewkes A, Warm PA (2000) Method of modelling the performance of rainwater collection systems in the UK. Build Serv Eng Res Technol 2:257–265

Fewtrell L, Kay D (2007) Microbial quality of rainwater supplies in developed countries: a review. Urban Water J 3:253–260

Freerain (2017) Commercial rainwater harvesting projects. http://www.freerain.co.uk/commercial-case-study-residential.html. Accessed 10 Oct 2017

Geographical (2017) UK Land becoming more urban. http://www.geographical.co.uk/places/mapping/item/1165-uk-land-becoming-more-urban. Accessed 12 Oct 2017

Gould J, Nissen-Petersen E (1999) Rainwater catchment systems for domestic supply: design, construction and implementation. Intermediate Technology Publications, London

Greater London Authority (2015) London sustainable drainage action plan – draft for public consultation. Greater London Authority, London. https://www.london.gov.uk/sites/default/files/lsdap_final.pdf. Accessed 10 Oct 2017

Gwenzi W, Dunjana N, Pisa C et al (2015) Water quality and public health risks associated with roof rainwater harvesting systems for potable supply: review and perspectives. Sustainability Water Qual Ecol 6:107–118

Hackney Council (2007) The SuDS manual. CIRIA C697, Hackney Council, London

Hassell C (2005) Rainwater harvesting in the UK–a solution to increasing water shortages. In Proceedings of the 9th international conference on rainwater catchment cistern systems. Petrolina

Heggen RJ (2000) Rainwater catchment and the challenges of sustainable development. Water Sci Technol 42(1–2):141–145

Helmreich B, Horn H (2009) Opportunities in rainwater harvesting. Desalin 248(1–3):118–124

Herrmann T, Schmida U (1999) Rainwater utilisation in Germany: efficiency, dimensioning, hydraulic and environmental aspects. Urban Water 1(4):307–316

House of Lords (2006) House of lords science and technology committee, 8th report of session 2005–06. Water Management 1: 191-I. London, UK

Hunt DVL, Rogers CDF (2014) A benchmarking system for domestic water use. Sustainability 6:2993–3018

Hurlestones (2017) Water solutions: case studies. Available via https://www.hurlstones.net/south-west/our-solutions-for-you-south-west/case-studies-south-west/. Accessed 10 Oct 2017

Jenkins J, Palframan L, Pericli A (2016) Customer attitudes to water use and its conservation. Br J Environ Clim Change 6(3):170–178

Kellagher R, Maneiro Franco E (2007) Rainfall collection and use in developments: benefits for yield and stormwater control. WAND Briefing note 19: report SR677, HR Wallingford, UK

Lash D, Ward S, Kershaw T et al (2014) Robust rainwater harvesting: probabilistic tank sizing for climate change adaptation. J Water Clim Change 5(4):526–539

Leggett DJ, Shaffer P (2002) Buildings that save water—rainwater and greywater use. Proc Inst Civil Eng Munic Eng 151(3):189–196

Leggett DJ, Brown R, Brewer D (2001) Rainwater and grey water use in buildings: best practice guidance. The Construction Industry Research and Information Association, London

Li Z, Boyle F, Reynolds A (2010) Rainwater harvesting and greywater treatment systems for domestic application in Ireland. Desalin 260:1–8

London Assembly (2016) London sustainable drainage action plan. https://www.london.gov.uk/WHAT-WE-DO/environment/environment-publications/london-sustainable-drainage-action-plan. Accessed 10 Oct 2017

Melville-Shreeve P, Ward S, Butler D (2016a) Rainwater harvesting typologies for UK houses: a multicriteria analysis of system configurations. Water 8(4):129

Melville-Shreeve P, Cadwalader O, Eisenstein W et al (2016b) Rainwater harvesting for drought management and stormwater control in the San Francisco Bay Area. In: Global performance of the drainage system – stormwater reuse, Novatech 2016, 9th International conference on planning and technologies for sustainable urban water management, Lyon 31 July–1 August 2016

Melville-Shreeve P, Ward S, Butler D (2016c) Dual-purpose rainwater harvesting system design. In: Charlesworth SM (ed) Sustainable surface water management: a handbook for SUDS 205. Wiley, London

Melville-Shreeve P, Hortsman C, Ward S (2016d) A laboratory study into a novel, retroffitable rainwater harvesting system. Br J Environ Clim Change 6(2):128–137

Memon FA, Lobban A, Butler D (2009) Modelling the effectiveness of rainwater harvesting as a stormwater management option. In: International association for hydro-environment engineering and research 33rd Congress, Vancouver, 9–14 August 2009

Nouh M, Al-Shamsy K (2001) Sustainable solutions for urban drainage problems in arid and semi-arid climates. Technical documents in hydrology, UNESCO IHP-V vol 40, UNESCO, Paris, pp106

ONS – Office for National Statistics (2017) Overview of the UK population. March 2017. https://www.ons.gov.uk/peoplepopulationandcommunity/populationandmigration/populationestimates/articles/overviewoftheukpopulation/mar2017. Accessed 25 Oct 2017
Parkes C, Kershaw H, Hart J, Sibille R, Grant Z (2010) Energy and carbon implications of rainwater harvesting and greywater recycling. Report: SC090018. Environment Agency. Bristol, UK
Parsons D, Goodhew S, Fewkes A et al (2010) The perceived barriers to the inclusion of rainwater harvesting systems by UK house building companies. Urban Water J 7(4):257–265
Pericli A, Jenkins J (2015) Review of current knowledge: smart meters and domestic water usage. Marlow Foundation for Water Research. Commissioned report
Rainwater harvesting Systems (2016) Rainwater harvesting case studies. http://rainharvesting.co.uk/case_studies/. Accessed 10 Oct 2017
Roebuck RM, Oltean-Dumbrava C, Tait S (2011) Whole life cost performance of domestic rainwater harvesting systems in the United Kingdom. Water and Environ J 25(3):355–365
Spillett PB, Evans SG, Colquhoun K (2005) International perspective on BMPs/SUDS: UK—sustainable stormwater management in the UK. In: Impacts of global climate change, American Society of Civil Engineers, Reston, US, 15–19 May 2005
Sponge (2005) The gaps in the existing case for building sustainable homes to encourage sustainable lifestyles. Sponge, London
Sustainable Buildings Task Group (2004) Better buildings – better lives. Sustainable Buildings Task Group, UK
Trautner HJ (2001) Rainwater utilization at the Frankfurt Airport. World Pumps 414:44
UK Climate Change Act (2008) Chapter 27. The Stationary Office, UK
UK Flood and Water Management Act (2010) HMSO, London, UK
U.S. Department of Energy (USDE) (2006) Energy demands on water resources: report to congress on the interdependency of energy and water. USA
Ward S (2010) Rainwater harvesting in the UK: a strategic framework to enable transition from novel to mainstream. PhD thesis. Exeter University. https://ore.exeter.ac.uk/repository/handle/10036/106575. Accessed 5 July 2016
Ward S, Butler D, Memon FA (2008) A pilot study into attitudes towards and perceptions of rainwater harvesting in the UK. BHS 10th national hydrology symposium, Exeter. https://ore.exeter.ac.uk/repository/handle/10036/4251. Accessed 5 July 2016
Ward S, Memon FA, Butler D (2010) Rainwater harvesting: model-based design evaluation. Water Sci Tech 61(1):85–96
Ward SL, Butler D, Memon FA (2011) Benchmarking energy consumption and CO2 emissions from rainwater harvesting systems: an improved method by proxy. Water Environ J 26(2):184–190
Ward S, Memon FA, Butler D (2012a) Performance of a large building rainwater harvesting system. Water Res 46(16):5127–5134
Ward S, Barr S, Butler D et al (2012b) Rainwater harvesting in the UK: socio-technical theory and practice. Technol Forecast Soc Chang 79(7):1354–1361
Ward S, Barr S, Memon F et al (2013) Rainwater harvesting in the UK: exploring water-user perceptions. Urban Water J 10(2):112–126
Ward S, Dornelles F, Giacomo MH et al (2014) Incentivising and charging for rainwater harvesting – three international perspectives. In: Memon FA, Ward S (eds) Alternative water supply systems. International Water Association Publishing, London
Waterwise (2010) Water efficiency in new developments: a best practice guide. Water Wise, UK
Waterwise (2012) Water – the facts. Why do we need to think about water? Waterwise, London. http://www.waterwise.org.uk/data/resources/25/Water_factsheet_2012.pdf. Accessed 5 July 2016
Way CM, Martinson DB, Heslop SE et al (2010) Rainwater harvesting: environmentally beneficial for the UK? Water Sci Technol Water Supply 10(5):776–782
Wheater H, Evans E (2009) Land use, water management and future flood risk. Land Use Policy 26:S251–S264

Wijitkosum S, Sriburi T (2008) Impact of urban expansion on water demand: the case study of Nakhonrachasima city, Lam Ta Kong watershed. Nakhara: J Environ Des Plann 4:69–88

Williams K, Dair C (2006) What is stopping sustainable building in England? Barriers experienced by stakeholders in delivering sustainable developments. Sustain Dev 15:135–147

Woods-Ballard S, Kellagher R, Martin P et al (2007) The SuDS manual. Report no: C697, CIRIA, London. https://www.hackney.gov.uk/Assets/Documents/The-SuDS-Manual-C697.pdf. Accessed 10 Oct 2017

Chapter 7
Urban Stormwater and Flood Management

Kevin Spence, Jonathan Bridge, Duncan McLuckie, and Jaya Kandasamy

Abstract This chapter overviews floodplain and stormwater management from two separate jurisdictions that have different climate, geography and history though having a governance, heritage and culture that is similar. There appears to be an underlying challenge for urban flood and stormwater management to transform stormwater from a hazard to a resource and to control and absorb the effects of flooding through sophisticated, adaptable urban design, smart environmental monitoring infrastructure, land use planning, evacuation management and planning and early warning systems and educated, informed communities. The common obstacle, seen in this review of evidence from the UK and Australia, despite significantly differing environmental, historical and governance contexts, is the distributed nature of the problem and its possible solutions. Water suffers from a version of the 'tragedy of the commons' in which its position as a common good – or indeed, a common hazard – makes individual stakeholders reluctant or unable to participate in effective action to manage the whole system. However, the growing number of catchment partnerships and community-led flood management initiatives bring public and private stakeholders together with the water management problems they face and encourage them to take common ownership.

Keywords Floodplain management · tormwater · Policy · Practice comparison · Australia · United Kingdom

K. Spence · J. Bridge
Department of the Natural and Built Environment, Sheffield Hallam University, Sheffield, UK

D. McLuckie · J. Kandasamy (✉)
School of Civil and Environmental Engineering, University of Technology Sydney, Sydney, NSW, Australia
e-mail: Jaya.Kandasamy@uts.edu.au

V. Jegatheesan et al. (eds.), *Urban Stormwater and Flood Management*, Applied Environmental Science and Engineering for a Sustainable Future,
https://doi.org/10.1007/978-3-030-11818-1_7

Abbreviations

AAD	Annual Average Damage
AEP	Annual Exceedance Probability
AIDR	Australian Institute for Disaster Resilience
BTE	Bureau of Transport and Economics
CFMPs	Catchment Flood Management Plans
CSIRO	Commonwealth Scientific and Industrial Research Organisation
Defra	Department for Environment, Food and Rural Affairs
EU	European Union
FME	Floodplain Management Entity
FPL	Flood Planning Level
FRM	Floodplain Risk Management
FRR	Flood Risk Regulations (2009)
FWMA	The Flood and Water Management Act (2010)
LLFA	Lead Local Flood Authorities at local and regional scale
NFCERMS	National Flood and Coastal Erosion Risk Management Strategy for England
NFM	Natural Flood Management
NSW	New South Wales
RWH	Rainwater Harvesting
SuDS	Sustainable Urban Drainage Systems
UK	United Kingdom of Great Britain and Northern Ireland
WFD	Water Framework Directive

7.1 Introduction

Water in the UK is widely perceived as abundant, but this masks significant regional and socio-economic variations in water supply and consumption. UK annual rainfall varies from around 500 mm per annum in the south-east to more than 3000 mm per annum in the north-west of Scotland. This trend is inverse to regional variations in population, which vary from 13,871 people per square kilometre in London to fewer than 10 people per square kilometre in the Scottish Highlands (Office for National Statistics 2013). Differences in lifestyle, affluence, economic activity and infrastructure lead to significant variations in consumption, while the average per capita water demand across the UK is 150 L/day. In urban centres, it is typically greater than this, and some estimates suggest as much as 684 L/day is consumed in London, accounting for all uses (Fewkes 2012).

Overall demand, particularly in the south-east of England, has been exacerbated by increases in population, which are predicted to continue. Although the UK population is considerably less concentrated in major conurbations, e.g. 35.7% in 2014 (Defra 2018) than that in Australia, the population in the 64 largest urban areas in the UK is predicted to increase by up to 10 million people by 2062 (Champion 2014).

The major summer flooding in 2007, which affected hundreds of thousands of people over a wide geographical area, significantly altered UK public discourse and governmental prioritisation of stormwater. Even as legislative and regulatory change in response to these floods was being enacted, further major winter flooding in 2013/2014 and 2015/2016 increased the average annual damages due to flooding in the UK to £1bn, with the costs of flooding in 2015/2016 alone exceeding £5bn (Miller and Hutchins 2017). The vulnerability of urban areas to flooding was emphasised by the very obvious role played by development encroaching on natural floodplains, together with the inability of existing drainage infrastructure to cope with the excessive runoff from impermeable urban surfaces.

It is worth noting that UK cities, with the marked exception of London, are small in terms of both population and area by global standards. Thus, fluvial conveyance of storm peaks from upstream catchments plays a significant role in UK flood risk (Guerriero et al. 2017), particularly where pre-twentieth-century urban expansion has incorporated smaller natural watercourses into the urban sewerage infrastructure by culverting or canalisation. This influence of the wider catchment on urban flooding has prompted a growing focus on catchment-scale water management as a means of flood mitigation downstream. Natural flood management (NFM) is now the focus of major government-funded research and demonstration project investment (Environment Agency 2018). NFM encourages systematic and distributed change in farming practice, habitat management and stream conservation in smaller headwaters to 'slow the flow', holding water on the land and in lower-order streams for longer in order to delay and reduce peak flows in higher-order rivers flowing through cities downstream.

The European Union (EU) Water Framework Directive (WFD) acts as the second major driver of stormwater management in the UK; the regulatory framework, if not the WFD itself, will likely remain in place after Brexit in 2019 (House of Lords 2017). At the core of the WFD is the requirement to work to achieve 'good' or 'high' quality for all water bodies, coupled with increasingly severe fines for private water utilities and companies responsible for handling and treating combined sewage, if found responsible for unmanaged discharges to receiving waters. This has driven significant investment in stormwater conveyance and treatment infrastructure aimed at minimising combined sewer overflows from inadequate existing systems.

The planning system provides the final piece in the jigsaw for current stormwater and flooding management in the UK and the only component to explicitly consider pluvial stormwater as an influence on the design of space and form in the built environment. Developers of both greenfield and previously developed land are now required to ensure that runoff from the site comes as close as possible to the equivalent greenfield runoff rate unless exceptional circumstances apply. Since the Flood and Water Management Act (2010), the planning system has promoted the use of Sustainable Urban Drainage Systems (SuDS) in order to achieve these goals, although in most of the UK this remains non-mandatory and widespread uptake faces a range of obstacles (Melville-Shreeve et al. 2017).

In Australia, the Great Dividing Range in Eastern Australia provides a natural separation of slower and longer rivers flowing west, from those faster and shorter but wider coastal rivers flowing east. In the mountain and coastal regions of Australia, flooding can happen rapidly with a warning of only a few hours in some cases. West of the 'divide', flooding of rivers in the vast flat areas of central and western New South Wales (NSW) and Queensland, as well as parts of Western Australia, may last for 1 or more weeks or even months on some occasions with flood warnings sometimes issued months in advance. Floods in these areas can lead to major loss of livestock and damage to crops as well as extensive damage to rural towns and road and rail links. In Northern Australia, the big floods occur in summer or early autumn in association with tropical cyclones or intense monsoonal depressions producing staggering quantities of rainfall (1000 mm in a few days).

In coastal areas, the ocean level can have a significant influence on flood levels whether the entrance to the ocean is open or closed. The condition of untrained entrances and, for closed entrances, the height of sand berms at the outlet to the ocean, is influenced by tidal and wave action.

Australian climate is influenced by board-scale ocean circulation. El Niño translates from Spanish as 'the boy-child'. Peruvian fisherman originally used the term – a reference to the Christ child – to describe the appearance, around Christmas, of a warm ocean current off the South American coast. Nowadays, the term El Niño refers to the extensive warming of the Central and Eastern Pacific that leads to a major shift in weather patterns across the Pacific. In Australia (particularly Eastern Australia), El Niño events are associated with an increased probability of drier conditions. La Niña translates from Spanish as 'the girl-child'. The term 'La Niña' has recently become the conventional meteorological label for the opposite of the better known El Niño. The term La Niña refers to the extensive cooling of the Central and Eastern Pacific Ocean. In Australia (particularly Eastern Australia), La Niña events are associated with increased probability of wetter conditions.

Flooding is often quite localised and therefore not as closely tied to broad-scale controls like the El Niño-Southern Oscillation phenomenon. However the La Niña years of 1916, 1917, 1950, 1954 through 1956 and 1973 through 1975 were accompanied by some of the worst and most widespread flooding this century. It can safely be said that, over much of Australia, flooding is more likely than usual during La Niña years and less likely in El Niño years.

Compared to the UK, European settlement has been relatively recent in Australia. Since 1788, there have been more than 2300 flood-related fatalities in Australia. Australia-wide, the overall death rate due to floods decreased from around 24 per 100,000 people per decade in the 1800s to 0.04 per 100,000 per decade during the 1990s and the first decade of the twenty-first century. Although the general trend has been for a reduction in flood fatalities, spikes in deaths still occur from time to time, as in 2011.

Floods cause significant amounts of damage. The annual average natural disaster relief costs of floods in Australia were $314 million in natural disaster declared areas between 1967 and 1999 (BTE 2001). These figures ignore any events where damage was less than $10 M and ignore the indirect costs of flooding. These costs also underestimate the cost of disasters with the broader cost of floods to the community and could be expected to at least double these figures. The total economic exposure of communities to flooding in Australia is in the order of $100 billion

(BTE 2001). It is also estimated that the 2011 Queensland floods temporarily depressed gross domestic product growth by up to 1%, which also provides evidence as to how damaging floods can be (Reserve Bank of Australia 2011). Flooding in South East Queensland in 2011 resulted in over $5.6 billion in damages and resulted in a special tax levy where some Australian taxpayers had to pay up to an additional $500 income tax in 2011/2012. Flood damages are significantly higher than those of any other type of natural disaster experienced at the time of publication, in Australia. Flood is on average the most costly national disaster.

7.2 Challenges in Urban Stormwater Management

In the UK, while the relationship between climate change and the water environment – quality, quantity, supply and hazard – has been widely assessed (e.g. Kundzewicz et al. 2014; Arnell et al. 2015), the additional impact of urbanisation on this relationship has only recently been addressed (Miller and Hutchins 2017). Significant increase in the urban population over the twentieth century have exacerbated both fluvial and pluvial flood risks with around 2.4 million and 3.8 million properties, respectively, now at risk (Defra 2014). Although the rate of development on high-risk floodplain sites decreased slightly post-2007, a total of 250,000 new homes were built on floodplains in the period 2001–2014, and implementation of protective measures to increase building resilience to flooding, whether in new-build or retrofit, remains low (Kovats and Osborn 2016).

The UK is widely dependent on urban infrastructure, including highways, sewerage, building designs and layout of urban spaces – which dates back to early twentieth-century or nineteenth-century urban expansion. This historic urban realm was not designed with management of stormwater in mind beyond the simple expedient of removing it to watercourses or treatment plants as quickly as possible. The combined impacts of population growth, infilling of urban green space (increased impermeable cover by up to 24% from 1991 to 2011 widely across UK urban areas; Warhurst et al. 2014) and increased water demand per capita now routinely exceed the capacity of these systems. In London, the celebrated Victorian combined sewer system engineered by Bazalgette was intended to overflow into the River Thames no more than four times a year; it currently overflows on average almost once a week, leading to the UK being found in breach of European regulations on untreated discharges to the environment (Dolowitz et al. 2018).

Climate change predictions (Jenkins et al. 2010) indicate overall warmer, wetter winters and hotter, drier summers in the UK, but trends are variable spatially and also depend on emission scenarios and confidence levels in the regional climate models used. An updated set of climate projections for the UK is expected in 2018. In respect of extreme storm events, there remains considerable debate about potential changes, with the spatial and temporal scale of modelling emerging as a key factor in assessing the intensity and frequency of the most extreme events pertinent to urban stormwater management (Miller and Hutchins 2017). The interface of potential increases in urban populations and rainfall is projected to cause 1.2 million more people and £351 million in assets exposed to pluvial flooding by the second half of this century (Miller and Hutchins 2017).

The UK clearly faces challenges from stormwater both in quantity and quality, which are exacerbated by demographic and environmental trends. The use of green infrastructure such as rainwater harvesting and sustainable urban drainage systems (SuDS), within programmes of integrated urban water management, has to date failed to develop a mainstream profile within urban development practice in the UK (Melville-Shreeve et al. 2017). This does not mean that there is a lack of awareness of such measures. There is also a continuously increasing body of best practice and applied case studies in the UK (e.g. http://www.susdrain.org). However, a major survey of UK industry professionals by the Chartered Institution of Water and Environmental Management in 2017 confirmed that SuDS remain perceived as more expensive and more difficult to implement than established 'conventional' drainage systems (Melville-Shreeve et al. 2017).

Rainwater harvesting (RWH), at the building scale, is increasingly well-established across Europe, supported by a well-developed commercial market in RWH technology worth 340 million Euros in 2016 (Melville-Shreeve et al. 2016). However, the UK housing stock, whether in retrofit or new build, lacks the larger garden areas and building spaces required to make RWH feasible and is dominated by institutions, both private and public, which tend to be conservative and lack the incentives, information and technical knowledge to drive innovation (Parsons et al. 2010).

In Australia, climate change is expected to have adverse impacts on sea levels as well as the intensities of flood-producing rainfall events, both of which may have significant influence on flood behaviour at specific locations. Climate change projections released in 2015 show simulated increases in the magnitude of the wettest annual daily total and the 5% wettest daily total across Australia (CSIRO and Bureau of Meteorology 2015).

The impacts of climate change and the associated ramifications on the vulnerability of floodplain risk management (FRM) mitigation options and development decisions can be significant and therefore cannot be ignored in decision-making today. McLuckie et al. (2010) provide examples of the ramifications of potential impacts. In terms of sea level rise, annual average damage (AAD) to a house built at the flood planning level (FPL) in an area where flood levels are directly controlled by ocean levels could increase by more than 1000%, due to a high sea level rise scenario by around 2100. In a town not influenced by sea level rise, a 30% increase in rainfall could increase AAD by 300%.

In recent years, residential flood insurance has become more generally available. Insurance products are developed based upon flood risk. This may effectively exclude insurance to the more frequently and severely flood affected houses, due to the cost of risk-based premiums. The inclusion of flood insurance can increase premiums by up to $10,000, beyond the financial capacity of many in the community. The Insurance Council of Australia advise that the better information that insurers have to price flood risk, the more realistic prices can be as they are more certain of the risk exposure. Without this information, but with an expectation that a risk exists, premiums are likely to be priced conservatively. The exposure of communities to flood risk is also likely to change with climate change. Annual insurance premiums are based on current conditions and therefore do not build in long-term climate change impacts. However, these premiums would increase as risks increase.

Jurisdiction over stormwater drainage and stormwater quality falls within the ambit of Local Government Councils, principally because in most cases they are the determining authority for land use planning and have a responsibility to manage risks to their communities. The relationship between flooding and drainage is clear. Stormwater quality with design requirements of less than 1 year return period (typically 3 months and thus at the opposite end of the return period spectrum compared to flooding) is often treated separately from flooding, and this is evident from policy and procedural documents. For flooding, the focus is on risk to life and property, and the exposure of the community to these flows is generally on the floodplain, outside of waterways and stormwater drains. This generally occurs in events rarer than the 1- in 10-year annual exceedance probability (AEP), with most interest in key design floods used for reducing exposure to flood risk, such as the 1% AEP event, and extreme events such as probable maximum floods (which may be used to examine management of risk to life).

7.3 Governance in Stormwater Management

The summer floods of 2007 in the UK, which affected large parts of the country, were instrumental in defining present-day UK policy on stormwater management. Recognising the severity of events described by emergency service leaders at the time as "the largest peacetime emergency we have seen" (Pitt 2008), the UK government commissioned an independent review of the events, responses and lessons to be learned. The Pitt Review was completed in 2008 and made 92 detailed recommendations for government at all levels covering flood prediction, mitigation, readiness, resilience and recovery. In the decade since its publication, these recommendations have directed activity at all levels of government and in related commercial sectors. The review (Pitt 2008) identified six key themes under a simple heading 'Lessons from 2007: what people need', which effectively form the policy basis for all subsequent stormwater and flood management activity in the UK:

- Reducing the risk of flooding and its impact
- Knowing where and when it will flood
- Being rescued and cared for during an emergency
- Maintaining power and water supplies and protecting essential services
- Better advice and help for people to protect their families and homes
- Staying healthy and speeding up recovery (Pitt 2008)

The emphasis here is very clear on defining stormwater flood management in terms of direct impacts on human welfare during and after flooding, setting in place an operative physical and administrative infrastructure, which is prepared for and responsive to future flooding events.

Stormwater management relating to water quality, particularly in respect of urban stormwater runoff to water bodies and discharges to receiving waters from combined sewer overflows, is governed by the European Water Framework Directive which was implemented in 2004 and requires European Member states to achieve good qualitative and quantitative status of all water bodies, including inshore marine waters, against a rolling programme of monitoring and assessment targets. The WFD integrates a diversity of prior legislation across a wide range of water and environmental protection issues.

Negotiations around the secession of the UK from the European Union (so-called Brexit) are due to be completed by 2019 at the time of writing. However, in the interim and initially afterwards, major EU legislation relevant to flood management, including the Water Framework Directive 2000 and the Floods Directive 2007, is expected to remain within UK law under the EU Withdrawal Bill 2017 (House of Lords 2017).

National Framework in England The central theme of the Pitt Review was the need for coordination of a previously fragmented hierarchy of responsibilities for flood management. The decade since Pitt has seen the formalisation of this coordination at national, regional and local levels. Devolution of government within the UK means that national legislation in England and Wales differs in detail from that in Scotland and Northern Ireland. Here, we focus principally on the legislative framework in England and Wales.

The Flood and Water Management Act (FWMA) of 2010 and the Flood Risk Regulations (FRR) of 2009 combine to form the primary legislative response to the Pitt Review and the EU Floods Directive. The FWMA addresses the need for coordination by bringing together numerous strands of legislation and policy covering public health, land drainage, water resources, reservoir management, civil contingencies, planning, environmental protection and research. The FRR were established separately only in response to differences in legislative timescales, and the two pieces of legislation together define the responsibilities and roles of authorities and partner organisations in the modern management of flooding.

In respect of flood management, the executive response to the implementation of the FWMA was the definition by the Department for Environment, Food and Rural Affairs (Defra) of a National Flood and Coastal Erosion Risk Management Strategy for England (NFCERMS) (Defra 2011). This is managed by the Environment Agency, which is the executive public body for environmental protection and enhancement in England (with powers having been devolved to a separate body in Wales in 2013). Within the NFCERMS, the Environment Agency is responsible for the development of Catchment Flood Management Plans (CFMPs) which aim to integrate consideration of stormwater and flood management across the hydrological cycle at the unit scale of natural hydrological processes, in other words, individual catchments (or watersheds). This approach benefits from the ability to integrate the management of upstream and downstream processes across a catchment so that authorities responsible for dealing with flooding downstream are placed alongside those responsible for planning, infrastructure and riparian management upstream. The catchment-scale approach furthermore places flood risk management alongside water quality management, which is also coordinated using a nested multi-scale catchment approach within the Water Framework Directive.

County and Local Framework in England The Environment Agency in England regulates the activity of Lead Local Flood Authorities (LLFA) at local and regional scale. LLFA responsibilities are delegated to unitary authorities at metropolitan (city) and county council level. Working with smaller district authorities, LLFAs are responsible for the definition of Surface Water Management Plans which are the primary mechanism for local flood risk management strategies and their integration with the planning system through Development Plans and land use planning control. LLFAs are responsible for coordinating the delivery of flood mitigation and preparedness between the range of public and private agencies responsible for land management, water resources, civil defence, development and infrastructure as well as private asset owners and the general public. It should be emphasised here that throughout England, water and sewerage infrastructure are owned and maintained by private companies, operating largely for-profit within the regulatory frameworks imposed by the Environment Agency and government bodies responsible for drinking water standards.

Development and the Planning Process The vision of the Pitt Review (Pitt 2008) was a simplified, streamlined system of surface water management that would allow a step-change in physical and institutional resilience to future flooding. At the strategic level, supported by substantial investment in environmental science research and technical development in hydrological modelling (Environment Agency 2018) and a now well-established organisational hierarchy focused around catchments/river basins for both flooding and water quality, this vision has gone some way to being realised despite repeated major flood events in the decade since 2007.

The urban fabric, i.e. the physical structure and infrastructure handling surface water runoff and mitigating flooding for vulnerable properties, remains resistant to change. While there are many factors involved, not least the relatively slow rate of renewal and replacement within the built environment, a key limitation on the uptake of SuDS and other 'best practice' stormwater management implementation lies in the planning process. The Pitt Review vision carried through into the FWMA was a proposal to require LLFAs to institute a SuDS Approval Body with the express purpose of enforcing the inclusion of SuDS in major developments and promoting, approving, adopting and maintaining SuDS (FWMA 2010, Section 3). However, following consultation (Defra 2014), the implementation of Section 3 was placed on indefinite hold in England, and a system of SuDS approval within the existing, parallel National Planning Policy Framework was put in place instead (Ellis and Lundy 2016). This addressed a range of short-term, administrative and logistical concerns with the implementation of Section 3 but relegated the LLFAs and their flood risk management strategy to the role of a consultee within the planning process.

The priority of flood risk management and SuDS within the decision-making process during approval of development planning applications is therefore also reduced. SuDS, and the stormwater management benefits they bring, are recognised as desirable but economically proportionate and non-mandatory. A recent survey of SuDS delivery by construction and civil engineering practitioners in the UK revealed a strong desire for Government to take a lead on promoting green infrastructure through adoption of strong, mandatory legislative standards for development (Melville-Shreeve et al. 2017). Crucially, nearly 70% of respondents acknowledged that the reliance on the planning system to drive SuDS uptake was insufficient to achieve its goal. In this context, specific constraints (technical requirements, costs, space availability, unfavourable ground conditions) on SuDS implementation on individual sites were the principal reason for not meeting the guideline standards on SuDS in developments (Melville-Shreeve et al. 2017). Where SuDS were implemented, soakaways, permeable pavements and water butts (rainwater harvesting) were the most common on small sites, while geocellular storage, permeable pavements and ponds/detention basins were the most common interventions on large sites.

In Australia, floodplain management commenced early after its European occupation, with policy on flooding evolving at a different pace across Australia. Its beginnings can be seen in the early 1800s, in particular in the 1810 edict of Governor Lachlan Macquarie, which followed a series of fatal and damaging floods in the Hawkesbury-Nepean Valley in the western part of Greater Sydney. The edict assigned each settler whose farm was within the influence of known flooding an allotment on high land within a township for a dwelling, office, garden, storage and stockyard. Macquarie's intention was that settlers would live on the allotments, commuting to their actual farmlands to tend their animals and crops. A subsequent edit in 1817 indicated that those who followed it would obtain favourable consideration and protection from the government, while those who did not could be considered wilfully blind to their own risk and undeserving of future indulgences from the government.

Since about the 1850s, farmers in some areas built levees to keep floodwaters off their land, and some communities constructed levees and drains to exclude floodwaters and speed up drainage after heavy rains. These initiatives were not all effective, largely because there was a lack of appreciation of the range of potential flood severity for many years; attempts to manage floods were generally uncoordinated; and there was little understanding of the varying types of approaches that were best suited to particular environments.

In NSW, the 1950s marked a significant policy step in floodplain management. Severe flooding in the 1950s resulted in the construction of substantial flood mitigation works in eastern Australia, particularly in New South Wales. In addition, in the aftermath of the 1955 flood on the Hunter River, the NSW Government established a State-wide programme for the construction of structural mitigation works aimed at reducing the risk of riverine flooding to existing developments.

This programme involves a financial partnership between local, state and sometimes the Australian Government, through varying funding schemes. A formal review in the mid-1970s demonstrated that despite the expenditure of many millions of dollars by local, State and Commonwealth Governments to address existing flood risk, the cost of restoration, relief and assistance following floods continued to grow. Impacts on protected areas were reduced. Increases in costs were associated with additional development of flood-prone areas being approved with little or no consideration of flood risk, rather than because of any failure of the structural works.

This resulted in the implementation of stringent development controls that excluded development from large portions of flood-prone land. However, this resulted in significant community backlash due to the implications for property development and the accuracy of the associated maps. This resulted in a change of approach by the NSW Government to a more risk management-based approach outlined in the NSW Flood-Prone Land Policy. In first formulating the policy in 1984, the NSW Government had regard to two important facts: flood liable land is a valuable resource that should not be sterilised by unnecessarily precluding its development; and if all development applications were assessed according to rigid and prescriptive criteria, some appropriate proposals may be unreasonably disallowed or restricted, and equally quite inappropriate proposals may be approved. This risk-based approach is outlined in the NSW *Floodplain Development Manual*, which was first released in 1986.

While this strategy was successful in managing development within floodplains, it became increasingly apparent that, notwithstanding attempts to protect existing development and action to manage the growth in future flood damage through land use planning controls, a flood risk remained that needed to be addressed. This residual or continuing risk results from floods larger than those used to design protection works for existing development or on which land use planning controls are based for new development. The Nyngan (1990), Coffs Harbour (1996) and Wollongong (1998) events provided both evidence and experience in these matters. The role of emergency planning and management was strengthened to reduce this risk.

Since the 2000s, there has been increased focus on environmental issues and on taking a more holistic approach to floodplain management, as outlined in the update of the NSW *Floodplain Development Manual* in 2005. This involves consolidating a risk management approach that considers the impacts of the full range of floods up to, and including, the probable maximum flood; using different land use planning practices to limit the risk that will be created through the future development of floodplains; recognising, communicating and managing the residual risk that continues to exist where the protection provided by development controls and/or flood mitigation works are overwhelmed; developing more accurate and timely flood warning and emergency management capabilities and recovery planning to improve community responses to, and recovery from, flood disasters; considering cultural and environmental issues and community views when assessing flood mitigation and other flood risk management measures. It requires a coordinated multidisciplinary effort across all levels of government and between agencies and departments with different responsibilities and the support of non-government organisations and professionals in a wide range of industries. The outcome is advice to decision-makers on how to manage the risk of flooding to the existing and future community while considering community aspirations and supporting the built environment.

Using a strategic approach allows robust management plans and measures to be developed, which can consider changing risk due to influences such as better data, improved analysis methods, changing climate and intensification of development.

Such an approach supports sustainable management and long-term community resilience.

However, even today, flood risk management practice varies greatly around Australia, not just at a state or territory level but at regional and local levels, as those agencies with primary responsibility for managing flood risk, such as a Floodplain Management Entity (FME), are at different points on a path towards best practice. This variation occurs due to various factors, including societal, governance and resourcing priorities and the differing severity of flood risk across Australia.

National Framework in Australia The National framework was established in 1999 and updated in 2013 and 2017. The national framework provides a robust, fit-for-purpose approach to understanding flood behaviour and managing the full range of flood risk and connecting studies often undertaken at a community, locality and catchment scale to governance of flood risk that occurs within the service area of an FME. The best-practice approach promotes understanding flood behaviour (considering existing knowledge of flooding and current management practices), so that the flood risk to the community can be understood, effectively communicated and, where practical and justifiable, mitigated. It facilitates informed decisions on managing risk and economic investment in development and infrastructure on the floodplain. These features can be used to create a platform that works towards achieving the vision for best practice management outlined by the Australian Institute for Disaster Resilience (2017). The framework supports managing flood risk across an FME by:

- Supporting a risk management approach.
- Providing a basis for establishing, monitoring, maintaining and communicating the sustainable governance arrangements in which the FME manages its flood risk.
- Considering the profile of the community living in the floodplain. Community vulnerability and exposure to flooding may influence management decisions.
- Providing a structure for the FME to oversee flood risk management and access technical and policy advice from relevant government agencies.
- Providing the basis for collating, maintaining, using and sharing the best available information on flood risk and management, through a knowledge hub. The framework promotes the communication of this information within government to inform decision-makers in land use planning, flood risk management, flood forecasting and warning, emergency response and recovery management. It also provides the basis for communicating information to the community in a consistent format, which promotes improved community knowledge of, and resilience to, flooding.
- Outlining the importance of consulting the community to gather their knowledge of flood risk and obtain input on strategies to manage this risk.
- Providing a basis for monitoring and reviewing. Current knowledge and management of risk and assessing and prioritising efforts and resources to fill gaps in the short and long term.

- Linking floodplain-specific processes to management of flood risk at an FME level.
- Providing the basis for assessing treatment options and treating risk.

A cornerstone of the floodplain management process is the engagement of local flood affect communities. Community representation and engagement provides a gauge to what risk the local community is willing to bear in which locations and in what circumstances, which can then be translated into the development of floodplain management plans. These plans are bespoke for local flood-affected communities and outline the local FME's decisions on how they intend to manage their flood risk into the future. Funding for flood planning and flood mitigation for communities may have support from the State, and in some cases, Federal Governments, and requires a contribution from local communities through the FME (which is usually the local government council). In some ways this avoids local floodplain management becoming 'gold-plated' and results from a thoughtful, informed assessment of the flood risk a local community faces and the most effective means of reducing it within the resources that the community may be able to attract.

7.4 Knowledge Gaps and Barriers to Effective Stormwater and Flood Management

In the UK, site-scale flood and stormwater management is led by developers and coordinated to a degree by the local authority planning department, while action at larger scales is driven by LLFAs and the water utilities with the Environment Agency acting both as a guide/partner and as a regulator. This leads to the development of two distinct approaches to large-scale urban water management: a governance-led approach in which identification (or actualisation) of a significant flood risk prompts action to prevent losses and protect lives in the future and a utility-led approach, in which action is taken to reduce present and future liabilities within environmental regulation. Two case studies briefly illustrate the dimensions of these approaches and highlight the gaps in knowledge and institutional obstacles to effective implementation of management plans.

Governance-Led Flood Management Planning: Sheffield, South Yorkshire Much is made of the distinction between adaptive, 'resilience-based' approaches to flood management and mitigating, 'resistance-based' approaches. The former tends to be associated with ecological, 'natural' flood management (Environment Agency 2018) and a change in social attitudes and urban design strategy towards the water environment; the latter typically represents 'traditional' hard-engineered flood defence infrastructure (Goodchild et al. 2018). The development of flood management strategy in Sheffield, UK, following the major June 2007 floods, provides a lens through which to examine the tension between these approaches.

Nearly 10,000 Sheffield residential and non-residential properties lie in areas with an annual flood risk >0.1%, of which 4000 have an annual flood risk >1%. Following the initial clean-up after the 2007 flooding of the River Don and its tributaries, Sheffield City Council instituted a major Flood Protection Strategy involving a mixture of hard defences, river restoration and channel clearance, riparian storage (often allocating existing parks and amenity spaces) and softer measures such as incentives to upgrade properties at risk and work with businesses to improve response plans (Sheffield City Council 2013). Funding for smaller works was found from partnerships with Yorkshire Water (the local water utility) and the Environment Agency, while funding for larger-scale elements of the scheme is, at the time of writing, still being sought from the UK national government strategic flood defence funds.

In practice, while smaller-scale, engineering-based works have been implemented and have received widespread praise (e.g. 'daylighting' works to open up culverts and create 'pocket parks'), larger propositions such as the creation of riparian storage have met both negative public reaction and competition with other major cities for increasingly limited strategic funds. Barriers to successful, integrated and adaptive flood management at the city scale in Sheffield include the costs involved in innovation and consultation, the difficulties of engaging with small businesses and domestic property owners to encourage property-level measures, the lack of national support for transitional schemes, and the complexity of dealing with a multiplicity of stakeholders and landowners with particular interests (Goodchild et al. 2018). In many respects, a cynical public view attempts to transition to adaptive, resilient community-led flood management as little more than an excuse for the state to shift funding responsibility towards individuals.

Utility-Led Action on Stormwater Pollution: Thames Tideway Tunnel, London As noted earlier in this chapter, the Victorian-era sewage infrastructure of London is failing to adequately cope with increased combined sewage and stormwater loads resulting from population growth, increased water consumption and projected climate change. The unacceptable rate of raw discharge to the River Thames led to a breach of environmental standards as originally set out in the EU Urban Waste Water Treatment Directive (1991) and resulted in the commissioning in 2000 of a strategic study to find a solution (Dolowitz et al. 2018). It is important to note that the liability for these infringements falls on a private company, Thames Water Utilities Ltd. The incentive to act is therefore a direct corporate response to potential losses resulting from prosecution via the Environment Agency, and the costs of any response are transferred to water consumers (the general public within the supply region) via increases in water rates (Dolowitz et al. 2018; Morse 2017).

A hard-engineering solution, the Thames Tideway Tunnel with two major interceptor tunnels and upgraded sewage treatment and pumping facilities at an estimated cost of £4.2bn (Morse 2017) was quickly established as the preferred option. In 2011 a range of stakeholders impacted by the proposals set up a Thames Tunnel Commission to re-evaluate the alternatives, including more resilience-based green infrastructure solutions. Many of the issues raised in Sheffield were identified as problematic in the positioning of city-scale SuDS infrastructure in opposition to a single, large infrastructure project – the difficulties of working with multiple small stakeholders, the lack of coordinated planning and funding frameworks for green infrastructure, and the uncertainty in ability to predict the specific effects of SuDS, either individually or in combination. This last point is crucial. In the absence of specific regulatory standards for urban water quality, the target standards for any Thames Tideway solution were adopted around support for specific aquatic species and focused on river parameters such as dissolved oxygen concentrations, public health and aesthetic considerations. Against this yardstick, the diffuse nature, complex implementation and uncertain net outcomes of SuDS were judged to make these untenable in comparison with the Thames Tideway Tunnel. In its conclusions, the Thames Tunnel Commission regretted that, under existing governance frameworks, "it is easier to construct large, costly, inflexible and environmentally-impacting infrastructure systems, like the tunnel, than it is to provide green infrastructure alternatives that deliver many benefits to society and that are adaptable to a changing climate" (Dolowitz et al. 2018).

Both the Thames Tideway and the Sheffield case studies highlight the major challenges for urban flooding and stormwater management in the UK, if it is to make a widespread transition to sustainable, green infrastructure. The issues are threefold: lack of strong and coherent regulatory and legislative drivers; the difficulties and cost of engaging and working with diverse stakeholders in a complex social, political and economic landscape; and the current lack of a robust evidence base for the effects and benefits of green infrastructure, particularly in combination and over a range of spatial and temporal scales.

Embedding consideration of flood risk in land use planning in a strategic manner has been a major challenge in floodplain management in Australia. Land use planning that incorporates flood risk management is a very effective means of controlling flood risk. It needs to carefully consider emergency response planning as this may limit development potential, identify additional works required to appropriately manage flood risk in some cases or mean that development of the area would pose too great a risk. Some of the challenges of managing flood risk to new development on a strategic basis are that:

- *The danger to personal safety* posed by flooding should be minimised through land use planning by considering the compatibility of the development with the full range of flood risk. Emergency management needs to be an integral consideration in both developing planning instruments and in implementing these planning instruments through flood-related development control plans or policies and in rezoning and development decisions. Its consideration may result in exclusion of, or limitations on, development in some areas, but its neglect will leave some parts of the community exposed to unreasonable levels of danger to personal safety during a flood event.
- *The damage to private property* is to be managed to an acceptable level. This can generally be achieved by having development compatible with risk, through appropriate zoning and development controls, which may include FPLs and

associated minimum fill and floor levels. Appropriate siting and controls should increase the affordability of flood insurance.

- *Critical infrastructure availability* needs to be ensured, meaning it is available, accessible and capable of performing its intended function during an event. The range of expected conditions in which any new critical infrastructure is to operate and what functionality would be required should be considered and thus designed accordingly. This may result in design standards for some critical infrastructure that are significantly higher than general standards. For instance, flood evacuation access routes need to be designed so that these are capable of performing their role in the expected conditions. This may result in wider evacuation routes, larger return periods for the design of both, road and cross drainage, and an elevated road rising to high land further along the evacuation route or at an evacuation centre.
- *Damage to public infrastructure* is managed to an acceptable level. Life cycle costing of infrastructure, from an awareness of the full extent of flood risk, and the necessary repairs and downtime after a flood event, may lead to decisions to build more flood-resilient infrastructures.
- *Societal implications* due to flooding should be limited. Viability of the business sector can be an important issue and needs to be considered closely in managing flood risk to future development. Flood risk always has a cost. Either it is the up-front cost of protection provided by appropriate development controls, the cost of repeated repairs after flood events, the cost of site-specific mitigation works, increased council rates to pay for broader community management measures, or increased insurance premiums. Therefore, businesses should consider flood risk in their planning. However, governments need to consider this, setting appropriate development control for the business sector. Developers can make significant profits from lower protection levels, and government decision-makers may come under pressure to reduce standards. While the developer may subsequently make the savings, it is the owner or occupant who bears the ongoing cost of flooding, and this may put pressure on government to undertake works to provide increased flood protection to reduce damage and increase business viability. A reasonable level of protection should therefore be provided to the business sector. Another important issue that needs to be considered in assessing management options is whether sites have cultural and heritage significance: these need to be identified and not significantly adversely affected by any works.
- *Flood-dependant ecosystems* need to be protected or enhanced as part of any flood mitigation project. This may involve considering these in design by ensuring maintenance of connection with the waterway and an appropriate flood regime and ensuring that the land use zonings protect these areas.
- *Other natural hazard and environmental issues* are considered so that management of flood risk is compatible with management of these other concerns, such as stormwater management plans, maintenance of riverine corridors and long-term geomorphological changes to the river. These may influence the siting of

any flood mitigation works and in determining whether land should be developed, and if so, the type of development that would be appropriate and the conditions that should apply.

7.5 Conclusions

> Many issues are too intractable and too enmeshed in contradictory interests. We have problems, but we don't have the publics that go with them (Latour 2012).

Is this the underlying challenge for urban flood and stormwater management? It is clear that we have the technologies to transform stormwater from a hazard to a resource and to control and absorb the effects of flooding through sophisticated, adaptable urban design, smart environmental monitoring infrastructure, land use planning, evacuation management and planning and early warning systems, and educated, informed communities. The common obstacle, seen in this review of evidence from the UK and Australia despite significantly differing environmental, historical and governance contexts, is the distributed nature of the problem and its possible solutions. Water suffers from a version of the 'tragedy of the commons' in which its position as a common good – or indeed, a common hazard – makes individual stakeholders reluctant or unable to participate in effective action to manage the whole system.

The Thames Tideway Tunnel provides a stark example of where it is simpler for a small number of expert stakeholders to implement an expensive solution to a limited problem, passing on the cost to a resigned general public, than to engender widespread action on smaller, cheaper, distributed solutions with myriad wider benefits to the urban fabric, social and health outcomes and climate resilience. A crucial lesson for sustainable water management planning is the importance of the definition of desired outcomes in placing limitations on what constitutes an acceptable solution. Given the current state of knowledge and uncertainty in respect of large-scale green infrastructure implementation, overambitious or inflexible expectations (even when applied in good faith) can take innovative, experimental but potentially transformative solutions off the table.

The growing number of catchment partnerships and community-led flood management initiatives, which in the UK have flourished in the regulatory environment created by the Pitt Review, represents probably the most flexible and effective governance structures for sustainable community water management going forward into the significant environmental changes predicted over the twenty-first century. An essential objective is to bring public and private stakeholders together with the water management problems they face and encourage them to take common ownership. This is the path that has been taken in Australia.

References

Arnell NW, Halliday SJ, Battarbee RW et al (2015) The implications of climate change for the water environment in England. Prog Phys Geogr 39(1):93–120. https://doi.org/10.1177/0309133314560369

Australian Institute for Disaster Resilience (2017.) https://www.aidr.org.au/. Accessed 28 Sept 2017

BTE (2001) Economic costs of natural disasters in Australia. Report 103, Bureau of Transport Economics. https://bitre.gov.au/publications/2001/report_103.aspx. Accessed 28 Sept 2018

Champion T (2014) People in cities: the numbers. Future of cities: working paper. Foresight, Government Office for Science. https://www.gov.uk/government/uploads/system/uploads/attachment_data/file/321814/14-802-people-in-cities-numbers.pdf. Accessed 28 Sept 2018

CSIRO and Bureau of Meteorology (2015) Climate change in Australia information for Australia's natural resource management regions: technical report, CSIRO and Bureau of Meteorology, Australia. https://www.climatechangeinaustralia.gov.au/media/ccia/2.1.6/cms_page_media/168/CCIA_2015_NRM_TR_Front_Index.pdf. Accessed 28 Sept 2018

Defra (2011) Understanding the Risks, empowering communities, building resilience: the national flood and coastal erosion risk management strategy for England. The Stationary Office, London. https://assets.publishing.service.gov.uk/government/uploads/system/uploads/attachment_data/file/228898/9780108510366.pdf. Accessed 28 Sept 2018

Defra (2014) Delivering sustainable drainage systems. Department for Environment, Food and Rural Affairs; Department for Communities and Local Government, September 2014. https://assets.publishing.service.gov.uk/government/uploads/system/uploads/attachment_data/file/399995/RFI7086_sud_consult_doc_final.pdf. Accessed 28 Sept 2018

Defra (2018) Official statistics: rural population 2014/15. 22 Feb 2018. https://www.gov.uk/government/publications/rural-population-and-migration/rural-population-201415. Accessed 28 Sept 2018

Dolowitz DP, Bell S, Keeley M (2018) Retrofitting urban drainage infrastructure: green or grey? Urban Water J 15(1):83–91. https://doi.org/10.1080/1573062X.2017.1396352

Ellis JB, Lundy L (2016) Implementing sustainable drainage systems for urban surface water management within the regulatory framework in England and Wales. J Environ Manag 183(3):630–636. https://doi.org/10.1016/j.jenvman.2016.09.022

Environment Agency (2018) Working with natural processes: evidence directory. Project number SC150005. https://assets.publishing.service.gov.uk/government/uploads/system/uploads/attachment_data/file/681411/Working_with_natural_processes_evidence_directory.pdf. Accessed 28 Sept 2018

Fewkes A (2012) A review of rainwater harvesting in the UK. Struct Surv 30(2):174–194. https://doi.org/10.1108/02630801211228761

FWMA (2010) Flood and Water Management Act 2010, United Kingdom. https://www.legislation.gov.uk/ukpga/2010/29/contents. Accessed 31 Jan 2019

Goodchild B, Sharpe R, Hanson C (2018) Between resistance and resilience: a study of flood risk management in the Don catchment area (UK). J Environ Policy Plan 20(4):434–449. https://doi.org/10.1080/1523908X.2018.1433997

Guerriero SB, Glenis V, Dawson RJ et al (2017) Pluvial flooding in European cities – a continental approach to urban flood modelling. Water 9(4):296–313. https://doi.org/10.3390/w9040296

House of Lords (2017) Brexit: environment and climate change. European Union Committee, 12th Report of Session 2016–17. HL Paper 109. https://publications.parliament.uk/pa/ld201617/ldselect/ldeucom/109/10902.htm. Accessed 28 Sept 2018

Jenkins G, Murphy J, Sexton D et al (2010) UK climate projections: Briefing report. http://ukclimateprojections.metoffice.gov.uk/media.jsp?mediaid=87867. Accessed 28 Sept 2018

Kovats RS, Osborn D (2016) UK climate change risk assessment evidence report: chapter 5, people and the built environment. Contributing authors: Humphrey K, Thompson D, Johns D et al. Adaptation Sub-Committee of the Commission for Climate Change, London pp 1–168. https://

www.theccc.org.uk/wp-content/uploads/2016/07/UK-CCRA-2017-Chapter-5-People-and-the-built-environment.pdf. Accessed 28 Sept 2018
Kundzewicz ZW, Kanae S, Seneviratne SI et al (2014) Flood risk and climate change: global and regional perspectives. Hydrol Sci J 59(1):1–28. https://doi.org/10.1080/02626667.2013.857411
Latour B (2012) Waiting for Gaia: composing the common world through arts and politics. Equilibri 16(3):515–538. https://doi.org/10.1406/38755
McLuckie D, Babister M, Dewar R (2010) Adaptation for the impacts of climate change on flood risk. 50th Floodplain Management Authorities Conference, Gosford NSW, 23–26 February 2010. http://www.floods.org.au/client_images/1813464.pdf
Melville-Shreeve P, Ward S, Butler D (2016) Rainwater harvesting typologies for UK houses: A multi criteria analysis of system configurations. Water 8(4):129. https://doi.org/10.3390/w8040129
Melville-Shreeve P, Cotterill S, Grant L et al (2017) State of SuDS delivery in the United Kingdom. Water Environ J 32:9–16. https://doi.org/10.1111/wej.12283
Miller JD, Hutchins M (2017) The impacts of urbanisation and climate change on urban flooding and urban water quality: a review of the evidence concerning the United Kingdom. J Hydrol Reg Stud 12:345–362. https://doi.org/10.1016/j.ejrh.2017.06.006
Morse A (2017) Review of the thames tideway tunnel. Report HC 783, Session 2016–17, 2 Mar 2017, National Audit Office. https://www.nao.org.uk/report/review-of-the-thames-tideway-tunnel/. Accessed 28 Sept 2018
Office for National Statistics (2013) 2011 Census: population and household estimates for the United Kingdom, March 2011. Statistical bulletin, Office for National Statistics, 21 Mar 2013. https://www.ons.gov.uk/peoplepopulationandcommunity/populationandmigration/populationestimates/bulletins/populationandhouseholdestimatesfortheunitedkingdom/2011-03-21. Accessed 28 Sept 2018
Parsons D, Goodhew S, Fewkes A et al (2010) The perceived barriers to the inclusion of rainwater harvesting systems by UK house building companies. Urban Water J 7(4):257–265. https://doi.org/10.1080/1573062X.2010.500331
Pitt M (2008) The Pitt Review: learning lessons from the 2007 floods. Cabinet Office. http://webarchive.nationalarchives.gov.uk/20100702215619/http://archive.cabinetoffice.gov.uk/pittreview/thepittreview/final_report.html. Accessed 28 Sept 2018
Reserve Bank of Australia (2011) Statement on Monetary Policy. ISSN 1448–5133 (Print), ISSN 1448–5141 (Online), 3 February 2011.
Sheffield City Council (2013) Sheffield flood risk management strategy. https://www.sheffield.gov.uk/content/dam/sheffield/docs/public-health/floods/Sheffield%20Flood%20Risk%20Management%20Strategy.pdf. Accessed 28 Sept 2018
Warhurst JR, Parks KE, McCulloch L et al (2014) Front gardens to car parks: changes in garden permeability and effects on flood regulation. Sci Total Environ 485–486:329–339. https://doi.org/10.1016/j.scitotenv.2014.03.035

Chapter 8
Biodiversity and Ecosystem Services in Relation to the Management of Storm Water and the Mitigation of Floods

Lynn Crowe and Ian D. Rotherham

Abstract Water, biodiversity, and human wellbeing are intimately connected, and in urban centres, this is especially so. This chapter addresses issues of the relationships between ecology in terms of biodiversity and ecosystem services and the management of storm water and flood mitigation. The account firstly explains the development of key concepts and then considers contrasting case-study examples in the UK and in Australia. The impacts of storm water and flooding on biodiversity are considered in relation to the delivery of ecosystem services and potential future scenarios.

Keywords Biodiversity · Ecosystem services · Habitats · Landscapes · Nature · Human health and wellbeing

8.1 Biodiversity and Landscape Change

Biodiversity is a concept that emerged during the 1980s and is broadly taken to mean the overall richness and diversity of the ecological resource (Wilson 1988). The term may be applied at varying scales from the global or the continental to the national or the very local.

The influence of floods on ecological systems and their biodiversity varies with the nature, regularity, and predictability of the flood events. In many landscapes, floods and flooding are a part of the annual cycle, and in others, such as some Australian situations, they are long-term drivers of ecological systems adapted to periods of extended drought and then massive inundation (Rotherham 2008). In other situations, in which ecology is not adapted to floods, but where it is now increasingly subjected to inundations, the consequences can be adverse for many wildlife species and for vegetation (Rotherham 2015a, b).

L. Crowe (✉) · I. D. Rotherham
Department of the Natural and Built Environment, Sheffield Hallam University, Sheffield, UK
e-mail: L.Crowe@shu.ac.uk

V. Jegatheesan et al. (eds.), *Urban Stormwater and Flood Management*, Applied Environmental Science and Engineering for a Sustainable Future,
https://doi.org/10.1007/978-3-030-11818-1_8

The potential impacts of flooding and storm water on biodiversity are also hugely affected by the way the environmental parameters such as climate are changed and are changing and by the ways in which humans have changed landscapes. Some of the changes are long term and long standing, but many others have occurred dramatically in recent centuries or even in recent decades (e.g. Rotherham 2014). Clearly, the timing, nature, and scales of impacts differ radically between the UK and Australia. In the former case, entire ecosystems (especially low-lying river valley forests and wetlands) have been eradicated or transformed. Furthermore, much of the wider landscape has been turned over to intensive farming; upland areas have been drained, burned, and grazed; and increasingly there is overwhelming urbanisation. Climate change is also leading to more frequent and intensive flooding.

In Australia, the landscape has been shaped by long-term impacts of often extreme climate and the effects of human management of the land. The latter is in two main stages:

Firstly the native Aboriginal impacts through regular light burning of vegetation.

Secondly, since European settlement of Australia, large areas of land have been turned into European-style farming and stations with the grazing of domestic animals such as sheep and nonindigenous fauna such as cattle and grazing by nonindigenous fauna such as rabbits.

Also, and significantly, the routine light burning by native Aboriginals has been largely suppressed, and as a consequence, there have been major wildfires in vulnerable landscapes. Many zones of Australia experience extreme climates with long dry periods and short, irregular intensive precipitation; and current climate change scenarios are making this much worse.

Across the world, the removal of natural forest and grassland cover has caused and is increasingly causing soil erosion and degradation and significantly increased flood risk (e.g. Rotherham 2008, 2014). Both in the UK (very widely) and in relatively limited population centres in Australia, there is increased urban development, and it is here that many of the impacts of flooding and storm waters are most obvious.

Finally, for both case-study countries, the long-term loss or degradation of productive agricultural land poses a serious problem for future food security. In Australia, the tendency for soils to become salinated combines with erosion and droughts to threaten future production. In the UK, for example, the long-term decline of presently productive peat-based soils in drained former fenlands may drastically reduce food production within 50 years. This loss is compounded by expected sea-level rise and climate change scenarios.

8.2 Dysfunctional Ecosystems and Declining Biodiversity

It has become widely recognised in recent decades that human impacts on (a) land use and (b) climate are causing ecological systems to become dysfunctional and so failing to deliver the ecosystem services necessary for humanity (Rotherham 2014).

This is a widespread phenomenon and one that becomes very obvious in densely urbanised centres. Indeed, a whole range of functions and services delivered by nature are now severely compromised, and the associated problems are now better understood by planners and politicians. However, the basic issues are by no means new and under almost every major desert on the planet, there is evidence of a once-thriving civilisation that over-reached its resources. The overriding issue today is that human population, the scale of human impact, and rapid changes in climate are together pushing the consequences to critical levels, such as an unprecedented loss of biodiversity and a mass extinction of species across the planet (Thomas et al. 2004). The scale of human-driven transformations in the environment has led to the recognition today of a new geological era – the Anthropocene (Steffen et al. 2007) – in which humanity and not nature itself is the primary driver of change. The implications of such human impacts are only now becoming more widely recognised (Rotherham 2014; Thomas et al. 2004; Thomas 2013).

8.2.1 Natural Versus Modern Landscapes

A reality is that around the world, landscapes and their ecological functioning are not 'natural' as such but are 'eco-cultural', driven by intimate and long-term interactions between people and nature (Agnoletti and Rotherham 2015; Rotherham 2015a, b). Indeed, with a few remote or isolated exceptions, there is a human footprint in every ecological system in the world. However, it is clear that where landscapes are long-established and driven by long-term, predictable patterns of events in the environment, there is a substantial degree of effective ecological function (Rotherham 2017a, b, c). However, over time, as climate impacts and human influence increases, then ecosystem functions can change and, in some cases, begin to fail altogether. The latter is the case when modern intensive land-use systems and urbanisation displace those of the ancient landscapes (Rotherham 2014).

Furthermore, in the ancient systems, there is a significant degree of co-evolution of species and ecology to adapt to the landscapes in which they occur. As the systems begin to fail, then biodiversity is lost, and the ecology is changed. The ultimate situation is one where, as a consequence of these transformations, the delivery of ecosystem services to modern humanity is reduced or in some situations, eliminated (Rotherham 2015a, b).

8.2.2 Causes of Failure and Their Implications

The causes of failure are fairly simple to conceptualise though hard to quantify in practice (Rotherham 2014). Human-related changes and natural fluctuations are difficult to separate and in reality co-exist and act together often synergistically. Removal of natural vegetation cover changes the microclimate, and this may exaggerate other impacts. However, the most obvious effects of removal are increased speed of water runoff and enhanced sediment load due to surface erosion during storm events (Rotherham 2008a).

In urban areas the situation is further complicated by the combination of increased cover by hard surfaces (which again increases the speed and percentage of runoff), the closing of groundwater recharge points, and substantially reduced water quality due to eroded sediments and urban-associated pollutants such as oils and also physical debris and detritus. These issues were discussed in landmark reports such as the Pitt Report (Pitt, 2007) and by Hewett (2008).

With rural landscapes converted to intensive farming, the results include massive increases in the erosion of sediments and soils and pollution from chemicals such as herbicides and pesticides that are frequently applied broadscale to the countryside (Hynes 1963; Rotherham 2008a, 2014). Other increasingly significant impacts of change are through pollutants such as microplastics, which occur ubiquitously in contaminated runoff. Localised changes may be hugely dramatic in the cases of major industrial operations such as mining, open-cast mining, and quarrying (Rotherham 2008b).

Human impacts have increasingly severe consequences due to the factors and trends described above but also because of the removal of the baseline ecosystems adapted to flooding and their replacement with ecological systems of novel, recombinant communities, and limited functionality. The implications of such transformations are yet to be fully considered, but awareness of the likely consequences is beginning to grow (e.g. Monbiot 2015).

The immediate effects of an ecosystem breakdown on biodiversity are changes in species composition with decreases in pollution-sensitive species and increases in those tolerant of pollution (Hynes 1963). Ecological communities adapted to flooding and sediment deposition and scouring will displace those unadapted to such conditions. Long periods of inundation may alter core habitat qualities such as nutrient availability but also critical factors like oxygen availability to fauna and to plant roots (Mason 2001). Furthermore, flooding incidents increase habitat disturbance and eutrophication and so provide opportunities for aggressive, invasive species, both native and exotic. Extreme floods cause major scouring of vulnerable banksides and can led to reworking and redeposition of sediments, which might include historically deposition-contaminated materials from past industry, as on the River Rother in North Derbyshire in the UK (Rotherham 2008b). This same process may trigger invasion by aggressive species along watercourses and into new habitats (Rotherham 2017a, b, c).

8.3 Urban Versus Rural Situations

Increasing urbanisation exacerbates both problems of rapid surface water runoff and of contamination too (Hynes 1963; Mason 2001). Urban landscapes are ecologically fragmented and include extensive hard and non-porous surfaces. Furthermore, urban drainage systems are prone to blockages and being overrun by excessive water, and these may also cause flooding. Urban and industrial storm waters are also likely to be heavily contaminated by pollutants and carry significant loads of chemically benign sediments that cause further problems through physical deposition. However, it should always be remembered that urban areas may receive floodwaters and storm waters from upstream rural catchments, and their pollution loads may in turn carry downstream into other rural areas. A major problem for rural catchments is the increasing eutrophication of surface waters and often of aquifers too (Mason 2001; Moss 2010).

Urban watercourses may have extensive stretches where the river or stream is culverted or canalised and almost always is decoupled from the functioning floodplain. Fragmentation of the urban watercourse may present mobility problems for species moving upstream or downstream and may limit recolonisation following any extinction incidents (Hynes 1963; Mason 2001; Moss 2010).

Rural environments may have significantly more 'natural' elements in their watercourses, though this is not necessarily the case. In the UK, for example, extensive lengths of rural watercourse are canalised and separated from their floodplains. The major contemporary influences on storm water and flooding in rural catchments are intensive farming in lowland areas, with high sediment loads and agrichemicals in runoff and drainage, and combinations of intensive grazing and burning of vegetation in upland zones (Rotherham 2008a, b, 2014). The situation in Australia is strongly influenced by the country's often drought- and heat-prone climate, which can limit agricultural intensification. However, where European agricultural and forestry systems have been imported and imposed into the Australian landscape, then severe problems from species extinction and displacement have followed. In fire-prone landscapes, major wildfires in recent decades are associated with extreme climatic conditions and with the suppression of long-standing cultural practices. Extreme flooding events and wildfires are essentially different aspects of the same problem and, combined, can result in substantial erosion of spoil and deposition of sediments downstream (Rotherham 2014). Furthermore, where native vegetation and fauna have been displaced by a wholly cultural ecology that is unadapted to periodic flooding or to fires, then further problems may ensue with a continuing breakdown of ecosystem functions.

8.4 Global Urbanisation

The trend towards urbanisation is continuing globally at an increasing rate, and from the early 2000s, the world's human population for the first time became more than 50% urban (Rotherham 2014). A complicating factor in terms of subsequent impacts is that many of the rural landscapes are then either abandoned or turned to intensive agri-industrial usage. These twin impacts of urban expansion and rural change combine to break down ecological functions and, for humans, the delivery of ecosystem services (Agnoletti and Rotherham 2015; Rotherham 2015a, b). In particular, in many regions of the world, the mitigation and amelioration of flood risk and storm water are becoming less effective, and the delivery of related ecosystem services are reduced (CIWEM 2001). The adverse impacts include a decline of important biodiversity but also other costs to society such as direct economic impacts of flood damage and the reduction of benefits such as leisure, tourism, recreation, and educational benefits. Harvesting of natural products such as important fisheries may be compromised, and in some cases, the closure of shipping lanes and navigations due to sediment deposition can be problematic and expensive to resolve. Additional problems include the blocking of waterways and the eutrophication of waterbodies due to pollution by nitrates from agricultural runoff. Associated contamination of groundwater aquifers can prove expensive and dangerous to health if drinking water is affected.

These problems occur increasingly at a global level and go hand in hand with loss of associated wetland and riverine habitats and their biodiversity. Upland reaches of major rivers may be important spawning grounds for economically important fisheries (such as salmon), and the estuarine waters, marshes, and flats are hugely important for coastal fish stocks. Furthermore, the coastal mudflats are ecological powerhouses for breeding, migrating, and overwintering birds. Yet it is into these important ecosystems that the consequences of inland flooding, pollution, and erosion are deposited. In many parts of the world, these same areas are under ongoing pressure from urban and industrial development.

All these issues, combined with human-influenced sea-level rise, put the ecosystems in these zones at serious risk of irreparable compromise. The services and benefits associated with these systems have been recognised increasingly at a global level and at a national level (e.g. Watson and Albon 2011a, b).

8.5 Emergence of the Ecosystem Services' Concept

One response to the breakdown of ecosystems and declining biodiversity described above has been the development of models attempting to identify the specific services and benefits that accrue to human society as a result of properly functioning natural systems. Scientists and policymakers have recognised the importance of identifying, defining, quantifying, and even financially valuing these functions in order to better protect and plan their future management. These functions and benefits have become known as ecosystem services. If we are trying to enhance the liveability of our cities, then our management responses to extreme weather conditions and subsequent development of infrastructure should also attempt to maximise all these ecosystem services.

The roots of the ecosystem services' concept are found in a project initiated by the United Nations at the end of the twentieth century (Millennium Ecosystem Assessment n.d.). As a result of work on international conventions such as the Convention on Biological Diversity (CBD) and the Convention to Combat Desertification (CCD), scientists and policymakers realised there was an unmet need for accurate assessments of the state of global ecosystems. Furthermore, it was increasingly clear that there was a need to value such ecosystem services in ways that could be reflected into planning processes.

The Millennium Ecosystem Assessment (MEA) project was established in 2001 and brought together the research of 1300 researchers from 95 countries, their first report being published in March 2005 (MEA 2005). The results confirmed that human activities have changed most ecosystems and threaten the Earth's ability to support future generations. The MEA went on to advocate an 'ecosystem approach' to achieve the sustainable use of the products and services on which human society depends.

An 'ecosystem approach' tries to integrate the management of land, water, and living resources and aims to reach a balance between three objectives: (1) conservation of biodiversity, (2) its sustainable use, and (3) equitable sharing of benefits arising from the utilisation of natural resources (JNCC n.d.). Its successful implementation relies on the application of 12 basic principles (see Table 8.1).

The Millennium Ecosystem Assessment project went on to state that 'Nature's goods and services are the ultimate foundations of life and health' (Millennium Ecosystem Assessment 2005, cited in Coutts 2016), thus establishing the concept of ecosystem services. The UK government's Department of Environment, Food, and Rural Affairs defines ecosystem services as 'the processes by which the environment produces resources utilised by humans such as clean air, water, food and

Table 8.1 Twelve ecosystem approach principles

The 'ecosystem approach' is a strategy that promotes conservation and sustainable use of natural resources in an equitable way through the integrated management of land, water, and living resources. It is the primary framework for action under the CBD and is comprised of 12 principles that are complimentary and interlinked:
1. Recognise objectives as society's choice – human rights, interests, and cultural diversity must be taken into account
2. Aim for decentralised management (i.e. subsidiarity) – balance local interests and wider public interests and encourage ownership and accountability
3. Consider the extended impacts or externalities – base economic valuation upon all ecosystem goods and services, and not simply the commodity value of extracted goods
4. Understand the economic context and aim to reduce market distortion
5. Prioritise ecosystem services – ecosystem functions and structures that supply services must be conserved
6. Recognise and respect ecosystem limits
7. Operate at an appropriate scale, spatially and temporally, at macro- and microscales
8. Manage for the long term, consider lagged effects
9. Accept change as inherent and inevitable – but make trade-offs clear and equitable
10. Balance use and preservation
11. Bring all knowledge to bear
12. Involve all relevant stakeholders to foster equity and inspire active participation in the stewardship of ecosystems

Adapted from JNCC (n.d.)

materials' (Defra n.d.). The Millennium Ecosystem Assessment went on to classify ecosystem services as follows:

- Supporting services: The services that are necessary for the production of all other ecosystem services, including soil formation, photosynthesis, primary production, nutrient cycling, and water cycling
- Provisioning services: The products obtained from ecosystems, including food, fibre, fuel, genetic resources, biochemicals, natural medicines, pharmaceuticals, ornamental resources, and freshwater
- Regulating services: The benefits obtained from the regulation of ecosystem processes, including air quality regulation, climate regulation, water regulation, erosion regulation, water purification, disease regulation, pest regulation, pollination, and natural hazard regulation
- Cultural services: The nonmaterial benefits people obtain from ecosystems through spiritual enrichment, cognitive development, reflection, recreation, and aesthetic experiences – thereby taking account of landscape values (Defra n.d.)

Clearly, storm water management and flood regulation are one of the regulating services shown in Table 8.2. Management interventions attempting to mitigate and adapt to these events can also be considered using the 12 ecosystem approach principles summarised in Table 8.1. Different responses will have varying impacts on the full range of ecosystem services. Some responses could have negative impacts, but some could also enhance ecosystem services if integrated and developed effectively. This chapter specifically focuses on biodiversity impacts and the cultural services summarised above. These include the spiritual, recreation, and aesthetic attributes of the natural environment, which can bring such benefits to human health and wellbeing, particularly in urban environments.

There is now good evidence about the wider value of green spaces and access to the natural environment for human health and wellbeing (see, e.g. Ward-Thompson et al. 2010; Gomez-Baggethun and Barton 2013; and Coutts 2016). William Bird's report for the Royal Society for the Protection of Birds (2004) explores the benefits to physical health in engaging with nature. Physical inactivity has serious effects on human health, and evidence suggests it may cost the UK economy more than £8 billion a year. Outdoor activities, particularly walking, offer a cheap and accessible route to better health for all and address many of today's pressing public health issues. The continued use of green space for physical activity is strongly linked to the quality of the landscape – in terms of beauty, diversity, and contact with nature. The report indicates that varied and wildlife-rich natural environments

Table 8.2 Ecosystem services

Provisioning services	Regulating services	Cultural services
For example, food, freshwater, fuel, wood, genetic resources	For example, climate regulation, disease regulation, flood regulation	For example, spiritual, recreation, aesthetic, inspirational
Supporting services those needed for the provision of the other services, e.g. soil formation, nutrient cycling, primary production		

Defra (n.d.)

with inspiring landscapes are most effective in promoting sociable walking and a healthier lifestyle.

As previously described, urbanisation and changing climates are leading to more frequent and increasingly severe issues of flooding and peak temperatures. This can also increase the severity of pollution episodes in cities around the world. Particular issues are being experienced in British cities with rising summer temperatures and growing air pollution problems (Hall et al. 2012). In Australia, long-term drought and rising peak summertime temperatures are causing increasing concern. Associated problems include air pollution and severe fire risk in vulnerable areas. These impacts can also be addressed through the lens of an ecosystem services approach.

As we better understand the wider importance of ecosystem services, interest in nature-based solutions to these problems is growing. Rather than a reliance on wholly engineered solutions within the built environment, urban green space is potentially able to provide a range of regulating ecosystem services including storm water attenuation, heat amelioration, and air purification (Davies et al. 2017). The extent to which these benefits are realised is largely dependent on how green spaces within urban areas are designed and managed and the core objectives when such spaces are planned. The availability of funding and the understanding of ecosystem service concepts within local governments (often the primary delivery agents of city infrastructure) are also all significant.

8.6 Green Infrastructure: Maximising Ecosystem Services

Many researchers and planners have now adopted a systems way of thinking about how the totality of the different components of the natural environment work together to provide the ecosystem services on which humans depend (Coutts 2016). This approach has helped to establish the concept of 'green infrastructure' (GI), defined as 'an interconnected network of greenspace that conserves natural ecosystem values and functions and provides associated benefits to human populations' (Benedict and McMahon 2002). This is not a new concept, and its history can be traced back to the early writing of landscape pioneers such as Frederick Law Olmstead in the mid-nineteenth century and the 'Garden City' movements of the early twentieth century (Fainstein and Campbell 2003).

> No single park, no matter how large and well designed, would provide the citizens with the beneficial influences of nature; instead parks need to be linked to one another and to surrounding residential neighbourhoods. Frederick Law Olmsted (as cited by Coutts 2016)

Urban parks, street trees, parkways, forests, community gardens, and the myriad of other forms of private and public greenspaces, taken together and considered as a system, are what constitute a community's green infrastructure (GI). The UK government agency with responsibility for nature in England, 'Natural England', produced detailed guidelines on the planning and delivery of GI in 2009. This emphasised that greenspace should provide a multifunctional, connected network delivering ecosystem services. In creating a typology of different types of GI and linking the concept closely to the delivery of public goods and services, Natural England reinforced that GI provision should be just as fundamental to good urban planning and development as the parallel concept of 'grey infrastructure' of roads, energy networks, water, and sewerage systems.

In 2014, the multinational construction company, Arup, produced a report on GI with an emphasis on the use of well-managed and interconnected, multifunctional greenspace for both sustainable urban drainage and flood mitigation, alongside a comprehensive range of other public benefits (see Table 8.3). This report echoes the work undertaken by government agencies elsewhere, and the promotion of this approach to achieve a more integrated and wider range of ecosystem services demonstrates that these concepts are now part of mainstream solutions.

The Arup (2014) report provides a range of examples from cities around the world where improvements to green infrastructure have been developed alongside measures to improve their climate resilience. These examples include development undertaken after the failure of the structural flood defences in New Orleans in the USA following Hurricane Katrina in 2005. The State of Louisiana and City of New Orleans have taken steps to increase the resilience of the city to sea-level rise, hurricanes, and flooding, with a clear shift from structural defences to more natural solutions utilising green and blue infrastructure inspired by the Dutch experience (National Urban Forestry Unit 2010).

The City of Portland's sustainable drainage system installed in Oregon (USA) is another example of climate change-resistant measures in action. This system used the terraces of Mount Tabor for storm water processing, with the lower areas processing the water for human consumption. The system effectively copes with storm

Table 8.3 The benefits of green infrastructure

Environmental benefits	Economic benefits	Social benefits
Improved visual amenity	Increased property prices	Encouraging physical activity
Enhanced urban microclimate	Increased land values	Improving childhood development
Improved air quality	Faster property sales	Improved mental health
Reduced flood risk	Encouraging inward investment	Faster hospital recovery rates
Better water quality	Reducing energy costs via microclimate regulation	Improved workplace productivity
Improved biodiversity	Improved chances of gaining planning permission	Increasing social cohesion
Reduced ambient noise	Improved tourism and recreation facilities	Reduction in crime
Reducing atmospheric CO_2	Lower healthcare costs	

Arup (2014)

events whilst using nature to clean road runoff, as well as providing walkways and promenades to enhance public recreation. As a result, overflows into local rivers have been cut by 35% (City of Portland Environmental Services 2010). This sustainable drainage system has led the way in demonstrating how natural drainage can be attractively integrated into the urban landscape, as well as providing a much wider range of public benefits for local citizens.

In Rotterdam, in the Netherlands, the Spangen district has almost no open water and a high proportion of paved surfaces within a dense urban environment. In order to build resilience to climate change, a series of attractive well-vegetated water squares have been proposed in the district. The central area of each square has been lowered and paved to act as a buffer for rainwater, allowing runoff to occasionally fill the central area to reduce the impact of rainfall and storm events on the city infrastructure (Pötz and Bleuzé 2012). A critical objective of the Rotterdam authorities was the need to create substantial additional water storage capacity within the city. A parking garage beneath Museumplein Square was created with 10,000m^3 of underground storage. This initiative alone provides 12% of the water storage capacity required for the city centre.

Arup (2014) emphasises that all these examples have required cross-collaborative authority agreements. The realisation of co-ordinated climate resilience proposals for the future will require utility companies, businesses, and private and public sectors to adopt a more collaborative approach with a view to providing longer-term benefits of current actions and proposals. The following section reflects on how these principles might be applied to the Australian and UK contexts.

8.7 Australian Versus UK Environments: Issues and Context

The two case-study countries, Australia and the UK, differ immensely in terms of their environments, their climatic extremes, their biodiversity and ecological systems, and their patterns of urbanisation and of agricultural land use. Furthermore, within each country there are major regional variations of weather patterns, ecology, land use, and human settlement.

8.7.1 A UK Perspective: Overview

One major difference of course is that the UK is far more densely populated than Australia and the transformation of the countryside has been far more long standing and extensive than in Australia. In the UK, entire ecological systems such as the once-extensive lowland fens have been almost entirely eradicated (Rotherham 2013).

In the UK it has been suggested and evidenced that extreme weather combined with landscape degradation has triggered numerous catastrophic flooding events in recent years. Examples include York (1998, 2000), Sheffield (Yorkshire) in 2007,

Cumbria in 2013, Hebden Bridge (Yorkshire) in 2013, and others. In view of the increasing concerns over these extreme events, the UK Environment Agency has been working on evaluations and potential solutions, and this resulted in a report 'Working with Natural Processes to reduce Flood Risk' (Environment Agency 2017). Some of the proposed ideas are already being implemented, though the solutions like the problems are long term.

Examination of the UK situation in recent years has implicated the drainage and intensive management of upland landscapes in triggering massive downstream flooding, and this has been particularly the case in the Pennines – a chain of hills running north-south through northern England. The suggestion is that removal of peat and the peat-forming sphagnum mosses has combined with land drainage and heather burning for sheep grazing and grouse shooting, transforming the upper catchment and its delivery of ecosystem services downstream. This transformation of the landscape began during the parliamentary enclosures of the 1700s and intensified through the twentieth century with public subsidies for sheep grazing and for drainage schemes. An additional factor was the damage to pollution-sensitive sphagnum mosses by a combination of industrial and domestic coal-burning air pollution (Rotherham 2014, 2017b).

The intense air pollution has been reduced, and steps are now being taken to restore ecological functioning to the upper catchments with the blocking of drains or 'grips' and the reconstruction and remediation of damaged peat bogs. Other approaches are also being considered around major catchments such as the River Don, though some responses still go down the route of engineering infrastructure rather than restoration of ecosystem functionality. Major projects include the 'Moors for the Future' (Peak District National Park, 2018) and the 'Yorkshire Peat Partnership' (YPP, website, 2018).

Downstream initiatives include the construction of off-channel flood storage areas for use in the event of major flood incidents. Sites such as the RSPBs Old Moor Wetlands complex in South Yorkshire's Dearne Valley are examples of where biodiversity restoration now delivers a range of ecosystem services – flood mitigation, access and community health, leisure and tourism, education, and more (Rotherham 2008a; Rotherham and Harrison 2009; Rotherham 2013).

8.7.2 Case Study: The River Don, South Yorkshire, England

The River Don (Fig. 8.1) flows through South Yorkshire and the East Riding of Yorkshire, England. The river and its catchment have been the subject of long-term studies, e.g. Griffiths et al. (1996), Rotherham and Harrison (2012), Rotherham (2010, 2011, 2017a, b, c), and Firth (1997). There are also strategic reports and visions for the catchment, e.g. Edwards and Winn (2006) and Environment Agency (2005). It rises in the Pennines and flows for 70 miles (110 km) eastwards, through the Don Valley, via Penistone, Sheffield, Rotherham, Mexborough, Conisbrough, Doncaster, and Stainforth. It originally joined the River Trent but was re-engineered

Fig. 8.1 Location map of the Don River, South Yorkshire, England

by Cornelius Vermuyden as the Dutch River in the 1620s and now joins the River Ouse at Goole.

The Don can be divided into sections by the different types of structures built to restrict its passage. The upper reaches, and those of several of its tributaries, are defined by dams built to provide a public water supply. The middle section contains many weirs, which were built to supply mills, foundries, and cutlers' wheels with water power, whilst the lower section contains weirs and locks, designed to maintain water levels for navigation. The Don's major tributaries are the Loxley, the Rivelin, the Sheaf, the Rother, and the Dearne.

Together with its main tributaries (Rivers Rother and Dearne), the Don forms a river system with a catchment of 714 square miles (1850 km^2). The catchment holds over 1.5 million people in mostly urban settlements. Underlying geology is mostly carboniferous with coal measures now associated with coal mining and resultant pollution of the river system. The headwaters rise on the moorlands of the Pennines, where the rocks are largely millstone grit, whilst the lower reaches pass through areas of alluvial and glacial material, up to 66 feet (20 m) thick over Magnesian Limestone and Sherwood Sandstone.

Headwaters The source of the river is up on the high moors of the Pennine Hills at around 300 metres altitude. The water springs from acidic, wet soils and particularly peats on base-poor geology. Here the landscape provides water-related services of retention and absorption mitigating both downstream flooding and drought. There is significant biodiversity and productive agriculture, mostly pastoral, and also sporting economic interests such as grouse shooting and recreational and leisure provision from walking to outdoor sports. The landscape provides a substantial carbon reservoir and enhances climate mitigation. However, there have been substantial problems with drainage of large areas of mire since the 1500s or earlier, intensive drainage since the 1800s associated with 'land improvement' (sheep grazing and grouse shooting), the historic removal of peat for fuel by local communities (mostly pre-1600), gross air pollution from industrial cities (causing loss of sphagnum mosses and breakup of the peat mass with massive release of carbon), pollution of water supplies with brown discolouration, loss of water-related services, and consequent downstream flooding. The widespread burning of grouse moor vegetation causes loss of soil and soil nutrients and rapid runoff of storm waters with resultant flooding downstream. Current projects such as 'Moors for the Future' (Moors for the Future Partnership n.d.) are taking steps to restore the vegetation (sphagnum) cover to heal the pollution-related wounds and to block the drainage 'grips' and so 're-wet' the landscape.

Upper Reaches This zone has wet, acidic soils converted from upland peats to agricultural pasture land, forestry plantations, and some woodland. The area provides water supply for drinking water, recreation and leisure uses, productive farming, and limited carbon sequestration. Historic problems have included conversion of land to farming use with parliamentary enclosures from around the 1700s. To improve water supplies, there was some construction of weirs in the watercourse with canalisation and straightening of sections of the channel and drainage of adjacent marshes and bogs and the impoundment for reservoirs in the 1800s and 1900s. Removal of tree cover and conversion to farming have increased water runoff, and there has been consequent farm-related pollution such as nitrate enhancement. The major water supply reservoirs have substantially altered the natural functioning of the river system, which now relies significantly on compensation water. It may be possible to revert some adjacent pastureland to woodland and thus to 'slow the flow' of storm water runoff.

Middle Reaches (1) This is where the river runs through the towns and cities of Penistone, Stocksbridge, Sheffield, and Rotherham to Doncaster. The coal measures geology makes for varied soils with base-poor sandstones and base-rich shales. The region has extensive ancient woodlands, modern plantations, mixed farming, a history of mining, quarrying and industry, and increasingly urban residential use and associated transport infrastructure. In the east, the river cuts through the Magnesian Limestone at the Don Valley Gorge before opening out onto the great plain of the Humberhead Levels. The river is associated with water supply for drinking water and historically with water power for industry (until the mid-1900s), but there is now increasing recreation and leisure provision including riverside walking, wildlife watching, and angling. Away from the urban areas, there is productive agriculture and limited carbon sequestration. The river is increasingly important for biodiversity, and there are associated educational and health benefits of the watercourse. Historically, many of the tributaries were used to supply water for water power and were lined with off-channel dams. Associated with this were numerous weirs to control the water levels and lead water into the dams to turn waterwheels.

The result of the weirs was devastating to migrating fish such as salmon, trout, and eel. Through most of its length, the river is now canalised, and this is often in brick or concrete walls. Some tributaries are substantially culverted. The river is totally severed from its floodplain and associated marshes.

Urbanisation and industrialisation generated catastrophic pollution (chemical, organic, and thermal) so that by the 1970s, many sections were biologically dead. Introduction of non-native species (both accidental escapes from gardens and deliberate releases) has radically altered the ecology into what is now a thoroughly 'recombinant' system. Almost all ecosystem services were lost by the 1970s. Water pollution has declined dramatically though some runoff from highways still causes problems, and historically polluted land remains a dormant issue. There are modest attempts to create new off-channel flood areas, but they are limited, and in 2007 the river caused devastation to urban centres through massive flooding. Some tributaries are being 'daylighted', but other sections are still actively culverted. There is limited control of some invasive non-native plants. There has been reintroduction of fish species for angling – though there is also a widespread presence of introduced non-native fish. The Environment Agency also undertook programmes of introduction of oxygenating waterweeds to stretches from which they had been made extinct. Riverside walks have been constructed and have reconnected people to the river, and old factories have been transformed to spaces for urban living overlooking the watercourse.

Middle Reaches (2) The River Don continues from Doncaster east to the Humber Gap, the eastern plain beyond Doncaster having been a great wetland until the late 1600s with around 1000 km^2 of floodland and meandering rivers. At its centre was a huge, raised, multi-domed mire, many miles in extent. The intact landscape clearly provided huge carbon sequestration and a massive carbon sink and the area provided abundant resources of fish, peat-fuel, wildfowl, reeds, withies (coppice willow), and water management. The modern landscape provides agricultural resources, but these are time-limited due to climate change, sea-level rise, and peat degradation. The core remaining peat areas, which are still several kilometres in extent, have major biodiversity resources. The area currently supports some water-based tourism along the canalised navigations, but otherwise this is relatively limited. Incremental drainage and land improvement plus peat cutting for fuel have reduced the overall extent of the wetlands, but the region was still largely intact until the mid-1600s. From that date, a major process of drainage and 'improvement' took place until by the 1900s, only the core peat bogs remained; the rest of the land was converted to increasingly intensive arable farming. Throughout the 1900s, peat removal intensified for horticultural purposes; until by the late 1900s, the peat domes were entirely destroyed with consequent loss of water management services and the release of significant amounts of carbon to the atmosphere. Intensive arable soils in the wider areas around the current National Nature Reserve of the peat bogs still lose large amounts of carbon through breakdown of peat soils.

From the 1600s, the main channel of the River Don was shifted from its natural southern confluence with Trent, north to the River Ouse and thus to the Humber Estuary. The once meandering rivers are now fixed and canalised within the landscape and effectively disconnected from the floodplains. The lower reaches of the river receive pollution from upstream (including treated sewage from the urban areas and historically the industrial pollution too) and excessive runoff of nutrients and pesticides from the intensive agriculture. A relatively large area of cutover peat bog was acquired by the UK government nature conservation agency during the 1980s and 1990s, and this is now a National Nature Reserve (The Humberhead Levels). There are modest efforts now to reinstate some areas of riverine habitats, and a major project at Potteric Carr Nature Reserve has created new wetlands and reedbeds that provide water cleaning and flood mitigation services associated with nearby urban development. The Humberhead Levels restoration project seeks to re-wet the core peat areas of the former raised mires and, at 2887 hectares, is one of the biggest wetland restoration projects in Western Europe. However, beyond the boundary of the restoration site, there is an abrupt change into the fen-peat agricultural soils of what is still intensive farming. For topographic and climatic reasons, the farming here requires drainage to remove surface water and irrigation because of low rainfall. The soils are degrading, the system is unsustainable, and above all, the great peatland remains totally disconnected from the river system.

Lowest Reaches Having joined with the River Dearne at Denaby, the River Don flows north-east past Doncaster and then joins the Ouse at Goole after a course of about 70 miles (110 km). The combined river merges with the River Trent at Trent Falls, near the village of Faxfleet, to form the Humber Estuary. The River Don at the lowest reaches is totally separated from its ancient floodplain, and as the river grows, it merges with tributary streams to eventually become the Humber Estuary. The lower portions of the rivers here are or were historically tidal and with some saline influence. For much of this area, the highly modified river passes through the lower middle reaches to the Ouse confluence. The lower reaches have importance for biodiversity, for fish stocks, and for recreational and commercial river traffic. The whole system has been affected by pollution from upstream and industry around the lower estuary. Areas around the lower river have been urbanised landscapes or converted to intensive agriculture. Weirs, locks, and flood-control infrastructure have affected water behaviour and ecosystem functions such as fish migration. A Yorkshire Wildlife Trust project seeks to deliver benefits for wildlife and people through biodiversity enhancement, improved aquatic environment, community engagement and education, and physical access improvement.

The River Don and the Urban Communities Along its 100+ km, the River Don connects major and minor urban settlements, acting as both a physical and conceptual 'artery' through the landscape (refer to Table 8.4). From the headwaters on the high moors, the river provides water supply and numerous other services as described. Lower down the catchment, as described, the river provides both power and also a 'free' disposal service for the waste products of heavy industry. Historic

Table 8.4 Summary table of the River Don and its service provision

Section of the river	Characteristics	Ecosystem services delivered	Historic problems	Possible solutions
Headwaters	Upland peat-dominated catchment	Water supply, C-sequestration, biodiversity, recreation, agriculture	Drainage, air pollution, overgrazing, burning, agricultural conversion	Restoration of peat bogs, control of air pollution, reduction of grazing, control of fires
Upper reaches	Pastoral farmland and some woodland	Water supply, biodiversity, agriculture, tourism	Drainage, air pollution, overgrazing, agricultural conversion	Possibly tree planting, less intensive farming
Middle reaches (1)	Rural areas with woodland and farmland but more extensive urban zones	Water supply, biodiversity, recreation, health and wellbeing	Drainage, air pollution, water pollution, canalisation, culverting Groundwater lowering unsustainably with nutrient enrichment and other problems	Control of pollution, daylighting of watercourses, improved access to riverside, species reintroductions
Middle reaches (2)	Rural areas with extensive fenland and bog converted to intensive arable farmland, some urban areas, some industry, major transport infrastructures	Farming, biodiversity, recreation, health and wellbeing, C-sequestration, irrigation water	Drainage and conversion to intensive agriculture, air pollution, water pollution, canalisation, culverting Groundwater lowering unsustainably with nutrient enrichment, and other problems such as possible salination from seawater moving in	Re-wetting of extensive cutover peat bog, control of pollution, daylighting of watercourses, improved access to riverside, species reintroductions
The lowest reaches	Rural areas with extensive marshland converted to intensive arable farmland, some urban areas, some industry, major transport infrastructures	Farming, biodiversity, water-based transport, recreation, health and wellbeing, C-sequestration,	Drainage and conversion to intensive agriculture, air pollution, water pollution, canalisation, culverting, reclamation of land for urbanisation and industry, industrial pollution	Control of pollution, improved access to riverside, species reintroductions, restoration of floodland, development of tourism

changes all along the river have compromised the basic ecosystem functions of flood and drought mitigation, and in the urban reaches, the ecological processes had totally broken down by the 1970s. Today however, the riverside is more important for its provision of biodiversity and contact with nature, of health and wellbeing, and education. There is a growing recognition that the wider rural catchment and the urban river and its ecosystem services are intimately connected.

8.7.3 *An Australian Perspective: Overview*

With its extremes of climate and its soils vulnerable to erosion and to problems such as salination, Australia experiences major environmental issues and a significant loss of ecosystem service provision. Furthermore, the Australian landscape is altogether more 'ancient' than that of the UK (which is essentially around 10,000 years old and derived from the last glacial period), and the human footprint of the indigenous peoples has transformed and co-evolved the ecology over many tens of thousands of years.

The current scenario in Australian landscapes is the consequence of the more recent European colonisation, perhaps over 200 years, and of planetary-scale impacts of human-induced climate change.

The other major difference between the two areas under consideration is the limited urbanisation in Australia in comparison to the UK.

8.7.4 *Case Study: The Murray River, Australia (Based on the Report of the Murray–Darling Basin Authority (2011)* **The Living Murray Story – One of Australia's Largest River Restoration Projects***)*

At 2508 km in length, the Murray River is Australia's longest (Fig. 8.2). It rises in the Australian Alps and drains the western side of Australia's highest mountains before meandering across the extensive inland plains. Here it forms the border between the states of New South Wales and Victoria as it flows to the northwest into South Australia. Turning south at Morgan for its final 315 km, the Murray reaches the sea at Lake Alexandrina where it discharges into the Pacific Ocean. The river therefore breaks into three main sections: the headwaters in the mountains, the middle reaches across the extensive plains, and the final reaches as the river meets the sea. The river, especially in its lower reaches, has suffered major problems associated with the loss of water through extraction and usage. Consequent problems have included desiccation and landscape degradation, plus adverse impacts on breeding and migrating birds and on fish stocks. The historic timeline for the river following European colonisation does share some similarities of process with UK River Don, though clearly with a much foreshortened time period.

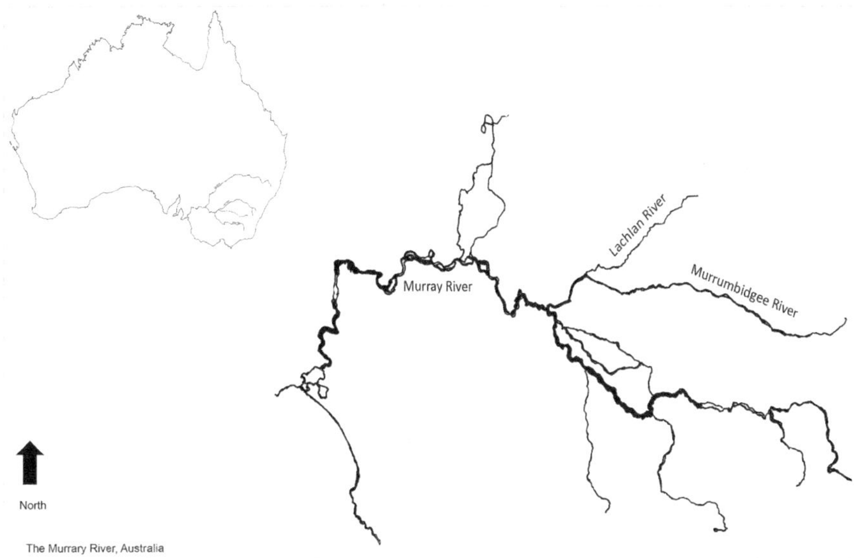

Fig. 8.2 Location map of the Murray River, Australia

8.7.5 A Timeline for the Murray River (from the Murray–Darling Basin Authority (2011))

The timeline presents key stages in the process from the massive deterioration in water quality and quantity and consequently in the ecosystem services delivered following the impacts of European settlement across the region. It then demonstrates the processes of recognition of the nature and scale of the problems and of the remediation works necessary to halt decline and to begin the recovery of a functioning river and its catchment. The timeline runs from 1850 up to 2011. Prior to the 1800s, the river was essentially a 'natural' watercourse in an 'unimproved' landscape but which had long-term human usage by native Aboriginals. The river would be subject to naturally varying inputs of rains and periods of natural long-term drought with ecological systems evolved and adapted to the specific conditions in this hot and often dry climate. From the 1800s onwards, the human impacts grew with European colonisation, associated agriculture, rural settlement, and some localised urban development. One consequence over the subsequent century or so was the collapse of the delivery of river-related ecosystem services, the nadir being reached from about 1980 to 1990.

1850: The first pumping schemes for the Murray River are installed.
1863: An intercolonial conference on navigation and management of the River Murray is held and agrees to make the major rivers navigable.
1887: There are irrigation settlements established at Renmark (SA) and Mildura (Vic).
1901: The Federation places constitutional powers relating to water resources in the hands of the individual states.

1902: The Interstate Royal Commission examines the conservation and distribution of waters of the Murray River.

1915: New South Wales, Victoria, and South Australia all sign the River Murray Waters Agreement, which effectively divides the water resources between them and establishes the River Murray Commission.

1922: The hydrological functioning of the river is modified by the imposition of a series of locks to control water flow and facilitate navigation. Lock 1 on the Murray River is completed in 1922, with ten more locks constructed by 1937.

1936: The Hume Dam is completed after 17 years of construction and further modifies the natural flow of the river.

1939: Barrages are constructed in South Australia to prevent seawater from entering the Lower Lakes.

However, over the subsequent decades, the processes of disruption and deterioration continue, and the river is drying up. Associated ecosystem services such as the provision of water, the facilitation of transport via navigation, and the support for fish stocks and breeding birds, for example, declined dramatically.

1981: For the first time in recorded history, the mouth of the Murray River closes due to the lack of freshwater flowing downstream. The associated problems are now too obvious to be ignored.

1987: The Murray–Darling Basin Agreement is signed, expanding the resource-sharing arrangements between the states to cover the whole Basin area. This establishes the Murray–Darling Basin Ministerial Council and provides for an increased focus on water quality.

1993: The Council (above) approves an annual Environmental Water Allocation of 100 GL to the Barmah–Millewa Forest.

1995: The Murray–Darling Basin Ministerial Council introduces 'the Cap', which amounts to a permanent restriction on the amount of water allowed to be extracted for consumptive uses each year from the Murray–Darling Basin.

1996: Queensland joins the Murray–Darling Basin Agreement, and the Australian Capital Territory also agrees to participate in the project.

1998: The Snowy Water Inquiry recommends environmental water release options to include the Murray River.

2002: The Murray–Darling Basin Ministerial Council proposes 'The Living Murray river restoration programme' and releases 'The Living Murray discussion paper'. This initiates basin-wide discussion about restoring the health of the Murray River system and begins the long-term process of recovery.

2003: After considering the outcomes of the community discussion process, the Murray–Darling Basin Ministerial Council announces 'The Living Murray First Step Decision' to begin returning the Murray River to the status of a healthy, working river with its associated ecosystem services. The fishway (or fish pass) at Lock 7 is completed.

2004: The 'Intergovernmental Agreement on Addressing Over-allocation and Achieving Environmental Objectives in the Murray–Darling Basin' formalises the agreement between partner governments to implement the First Step

Decision – notably the commitment of $500 million to recover 500 GL of water for six key 'icon' sites. The amount of $150 million is committed for water management structures to facilitate delivery of this water and the fishway at Lock 9 completed.

2005: The 'Living Murray Business Plan' is released in relation to the actions and milestones in the Intergovernmental Agreement; and the 'Living Murray Community Reference Group' is established.

2006: Funding commitment for The Living Murray water recovery is increased to $700 million and Works and Measures Program to $270 million. The 'Living Murray Indigenous Partnerships Program' is established after the signing of a memorandum of understanding with the Murray Lower Darling Rivers Indigenous Nations. The fishway at Lock 10 is completed. However, in this year, the recorded Murray River inflows are the lowest ever known, and problems are worse than ever.

2007: The Murray–Darling Basin Commission enters the water market for the first time to purchase irrigation entitlements. Nevertheless, because of the severe drought, only 22 GL of water is delivered to the key icon sites. Whilst very good localised environmental benefits are observed, the health of most key icon sites continues to decline. The Commonwealth Water Act 2007 is passed, establishing the Murray–Darling Basin Authority. This requires the Authority to develop a strategic plan for the integrated management of water resources across the Basin and establishes the Commonwealth Environmental Water Holder.

2008: The Water Act (as above) is amended to give effect to the Intergovernmental Agreement on Murray–Darling Basin reform. The Commonwealth Government purchases water for the first time. By June 2008, 17 GL of The Living Murray environmental water is delivered to key sites, and pumping begins from Lake Alexandrina to Lake Albert. This is in order to maintain water levels in Lake Albert and avoid the risk of acidification. Most key icon sites with floodplains or shallow waters are now dry or almost dry and support few waterbirds. The fishway at Lock 1 is completed.

2009: By June 2009, 343 GL (LTCE) of water is recovered for 'The Living Murray'. Then, 7 GL of The Living Murray environmental water is delivered to the key sites to protect threatened species and maintain important refuges during the continuing drought. An annual aerial survey of waterbird populations finds a 44% increase from the 2008 survey, but record low water levels in the Lower Lakes have led to high salinity levels and increased risk of acidification. A 16-member Basin Community Committee is established.

2010: By June 2010, 472 GL (LTCE) of water is recovered for 'The Living Murray', with 66 GL of 'Living Murray' environmental water delivered to the key icon sites. Environmental monitoring indicates that 79% of River Red Gum and Black Box communities at the key icon sites are in a stressed condition. Environmental works start at Gunbower Forest, Chowilla Floodplain, and Mulcra Island. Fishways at Locks 3, 5, and 6 are completed, and a guide to the proposed Basin Plan for integrated management of Basin water resources is released.

2011: By June 2011, 486 GL (LTCE) of water is recovered for 'The Living Murray', and 271 GL of 'The Living Murray' water is delivered to the key icon sites. This is the largest volume of water since the programme began. Environmental works start at Koondrook–Perricoota Forest. Encouragingly, Murray River summer inflows are the highest on record, and subsequent flooding results in major waterbird breeding events at Barmah–Millewa Forest and other sites along the Murray. The proposed Murray–Darling Basin Plan is released, and this timeline ends at the point of an apparent upturn for the Murray River and the recovery of at least some of its former ecosystem services.

Summary Essentially, following around 150 years of water abstraction and engineering manipulation (impoundment) and agricultural development combined with extreme weather events, the Murray River outfall to the sea had fallen catastrophically. Water levels and flow in the river were drastically reduced. Hand in hand with severe droughts were occasional, major flooding events. In effect, the ecosystem services provided by the river some 150 years earlier had now deteriorated, in some cases to the point of almost total loss. Faced with this major and growing environmental and economic catastrophe, by the 1990s, the regional governmental bodies and other key stakeholders began to formulate responses and actions. Then in 2004, the Australian government together with regional stakeholders made a commitment to the restoration of a functioning river basin. The major environmental issues faced by The Murray River are now being addressed, and the long-term process of recovery is underway.

8.7.6 A Comparison of the Two Situations

Whilst the situations in Australia and the UK might appear at first sight to have little in common, there are similarities in the social and economic drivers that have brought about landscape transformations and that today result in reduced ecosystem service delivery.

In the UK, a major driver of this transformation was the period of the parliamentary enclosures and 'land improvement' from the 1700s to the mid-1800s (Rotherham 2014, 2017b). Essentially, this period saw the removal of wetlands and the canalisation of rivers, plus the start of mass urbanisation and intensification of agriculture (Purseglove 1988). Rural populations were mostly removed from the land and either left to colonise countries like Australia and the USA or else sought employment in the expanding towns and cities. The same ideas and ethics were brought to the Australian continent following European migration. The activities were played out in a radically different and more extreme environment, but the processes were essentially the same (Rotherham 2014).

The end results are different because of the contrasting environmental conditions. In both cases, the baselines of climate and environmental conditions are becoming more extreme, and the greater extremes are becoming more frequent.

However, the UK tends to a more moderate and wetter climate but with farming that is more intensive and urban development that is more widespread and dominant in the landscape.

In both scenarios, the rivers' ecosystem services were disrupted by water abstraction for human consumption and for agriculture, by the construction of locks and/or weirs (for navigation in the former and industrial water power in the latter), and by widespread engineering works and modifications to the channels and banks. In each case, the rivers were affected by pollution from runoff, discharges, and soil erosion.

Australia is much hotter and drier but with occasional extreme events of droughts, catastrophic flooding, and extensive wildfires in vulnerable areas. The urban impacts in Australia are far more restricted but so too are the areas of land and climate suited to major urban developments. Australia, like the UK, also lost many of its extensive wetlands to the 'land improvers' of the nineteenth and twentieth centuries, and again this compromised key aspects of ecosystem services.

Overall, as noted above, there are key issues, themes, approaches, and consequences identifiable between and within the two environments. In terms of future trends, it is likely that the weather in the UK will become increasingly extreme with longer, deeper droughts and periods of intensive precipitation and consequent flooding. Indeed, with continuing pressure to urbanise, greenspace is increasingly threatened, and ecosystem service provision will be further compromised.

In Australia, it is likely that weather extremes will deepen and the events will continue to be intense and damaging. Key ecosystem services in terms of flooding and storm waters will be further disrupted. There will be continuing pressure on the limited productive agricultural lands because of obvious demand to produce food, and this will be exacerbated by problems such as extreme salination of many soils.

The other major difference between the two exemplar scenarios is the much greater extent of urbanisation, industrialisation, and associated pollution in the UK and the extreme climatic stresses in Australia.

8.8 Approaches and Mitigation Actions

In both case studies, it is clear that responses to the problems came about through increased awareness of the failure (and potential costs) of ecosystem services and the necessary policy or legislative frameworks. Funding of long-term restoration projects was then possible although it tends to be concentrated in limited areas. A key legislative trigger in the UK was the 'The Water Framework Directive' (EU 2000), adopted in 2000 and reviewed at various times, the latest being in 2015. This commits European Union member states to achieve good qualitative and quantitative status of all water bodies by 2015. However, the directive is adopted and implemented through national actions, and it was always obvious that the specific targets would not be reached. Nevertheless this does provide an aspirational framework for national and regional actions.

In the case of the River Don and the adjacent rivers, there are various actions either taking place or proposed, to reverse historic damage and to recover lost ecosystem service benefits. The range of measures proposed are a mixture of 'natural flood management', other environmental improvements, and new infrastructure including:

- Reduction or elimination of major pollution
- Daylighting of some culverted streams
- Restoration of some river meanders and 'natural' flood banks
- Creation of new woodland areas
- Re-wetting and restoration of damaged peatlands in the upper and lower reaches
- Creating new flood retention areas
- Improving certain flood defence and structures such as swales to slow the flow
- Clearing obstructions along the river (especially at pinch points like bridges), to help reduce water levels, and targeted lowering of the riverbed to increase capacity and flow
- Improving riverbank protection and installing debris dams
- Localised construction of raised defences with landscaping, terracing, embankments, and walls
- Creation of new wetland nature reserves, which also provide water cleaning and floodwater storage – these also provide biodiversity, education, and leisure facilities with health benefits

By contrast, for the Murray River, which is hugely significant for its biodiversity and also for the supply of water for agricultural irrigation and human consumption in towns throughout the catchment, a major issue is the recovering of the shortfall of water in the system. With the discharge to the ocean reduced (in many years to around one quarter of the natural flow and in some years to zero), there are consequent problems. Policy is led by the national Federal Government with responsibility for the catchment devolved to the states. Major issues include the need to discharge silt, salt, and pollutants out of the river system; and if the Murray Mouth closes, the toxins and nutrients accumulate to cause major problems. Engineering infrastructure along the river has also disrupted migrating fish stocks as indeed also happened in the River Don. In both case studies, the solution has been to strategically place fish passes at key points along the system. The Murray River catchment also has many cultural heritage sites throughout the area, and much of the river is significant to native Aboriginal peoples.

Water use, reuse, and recycling are important in the Murray River plan, and the River Don also has a Catchment Abstraction Plan to manage water use. Salinisation of irrigated agricultural land remains a risk for the Murray catchment and is a localised problem associated with lowered groundwater and seawater intrusion into the bedrock in the lower River Don.

In both examples, the wider landscape issues are addressed in part by extensive nature reserves and protected areas and newly restored lands. These locations are important in delivering biodiversity, educational, and health and wellbeing benefits.

In both situations, the remediation of ecosystem services such as provision of fish stocks has included better management of river flow, control of pollution, reversion of damaging engineering works, and the construction of fish passes or fishways to allow movement of fish stocks upstream.

8.9 Conclusions

This brief overview identifies some overall lessons, themes, emerging trends, and issues. Essentially there are three main drivers of the current breakdown in ecosystem service delivery:

1. Intensification of land use in recent centuries
2. Increasing urbanisation
3. Extreme climate change scenarios

Furthermore, with likely continuation of these trends associated with factors such as human population increase, more urbanisation, and further climate change, the problems will get worse in the decades to come. In both case-study regions, there are attempts by governments and other stakeholders to address some of the core issues. However, since 2007, the global economic downturn has affected key stakeholders in their ability to either produce viable strategic visions or, significantly, to deliver long-term implementation. The costs of remediation to resolve ecosystem services and the economic impacts of ecosystem failures – such as major flood incidents – are placed differentially in the national accounting systems. Societies therefore find it difficult to effectively target the necessary resources to the key actors able to initiate long-term solutions.

Additionally, factors such as land degradation, sea-level rise, global food insecurity, and climate change combine with human population rise to suggest that pressure for agricultural intensification will grow rather than reduce in the decades to come. In Australia, farming loss to salinisation will be a compounding factor, and in the UK, the 'breadbasket' areas of the lowland fens may have limited life expectancy as productive farmland, due to the degradation of peat-based soils and sea-level rise. It appears therefore that these ecosystem services, which are already at breaking point, may be further stressed.

The findings from this comparison are clearly transferable to elsewhere since these are global problems. A positive note is that we now have a good understanding of the processes at work and there are regional projects such as 'Moors for the Future' in the UK and the 'Murray River Project' in Australia that demonstrate the possible ways to address key problems. The ecosystem service's approach provides an overarching framework, in which the cost and benefits of actions (or inactions) can be placed and valued. There remain issues, however, in more fully embedding the initiatives into long-term policy and funding packages.

The recommendations for future work from this review include a wider collation of case studies from other countries and in a range of contrasting environments. In particular, it is increasingly important to gain a better understanding of where and when ecosystem services may fail in the future and the steps necessary to minimise damage or to avoid failures.

References

Agnoletti M, Rotherham ID (2015) Landscape and biocultural diversity. Biodivers Conserv 24:3155–3165

Arup (2014) Cities alive: rethinking green infrastructure. https://www.arup.com/publications/research/section/cities-alive-rethinking-green-infrastructure. Accessed 9 Mar 2018

Benedict MA, McMahon ET (2002) Green infrastructure: smart conservation for the 21st century. Renew Resour J 20(3):12–17

Bird W (2004) Natural fit – can green space and biodiversity increase levels of physical activity? Royal Society for the Protection of Birds, Sandy

City of Portland Environmental Services (2010) Mt. Tabor Invasive Plant Control and Revegetation Project. www.portlandoregon.gov/bes/article/316719. Accessed 9 Mar 2018

CIWEM (2001) Sustainability – development and flood risk. Proceedings of the CIWEM national conference 2001, CIWEM, London

Coutts C (2016) Green infrastructure and public health. Taylor and Francis, Wallingford

Davies HJ, Doick KJ, Hudson MD, Schreckenberg K (2017) Challenges for tree officers to enhance the provision of regulating ecosystem services from urban forests. Environ Res 156:97–107

Defra (n.d.) Ecosystem services – living within environmental limits. http://www.ecosystemservices.org.uk/ecoserv.htm. Accessed 1 Nov 2017

Edwards AMC, Winn PSJ (2006) The Humber Estuary, Eastern England: strategic planning of flood defences and habitats. Mar Pollut Bull 53(1–4):165–174

Environment Agency (2005) Planning for the rising tides. The Humber flood risk management strategy. Consultation Document August 2005. Environment Agency, Hull

Environment Agency (2017) Working with natural processes to reduce flood risk. https://www.gov.uk/government/publications/working-with-natural-processes-to-reduce-flood-risk. Accessed 31 Oct 2017

EU Council (2000) Directive 2000/60/EC of the European parliament and of the council of 23rd October 2000 – the European water framework directive. EU Council, Brussels

Fainstein S, Campbell S (2003) Readings in planning theory. Blackwell, Malden

Firth CJ (1997) 900 years of the Don fishery: Domesday to the dawn of the new millennium. Environment Agency, Leeds

Gomez-Baggethun E, Barton D (2013) Classifying and valuing ecosystem services for urban planning. Ecol Econ 86:235–245

Griffiths P, Simpson F, Rotherham ID (1996) Hydrology and water quality in the upper Don Catchment. Results of the research feasibility study and proposals for further research. Unpublished technical report. Sheffield Centre for Ecology and Environmental Management, Sheffield

Hall JM, Handley JF, Ennos R (2012) The potential of tree planting to climate-proof high density residential areas in Manchester, UK. Landsc Urban Plan 104(3–4):410–417

Hewett C (2008) Making space for water is key to adapting to inconvenient truth. Energy, resource, environmental and sustainable management, January/February 2008, 6–8

Hynes HBN (1963) The biology of polluted waters. Liverpool University Press, Liverpool

Joint Nature Conservation Committee (n.d.) Ecosystem approach. http://jncc.defra.gov.uk/default.aspx?page=6276. Accessed 1 Nov 2017

Mason CF (2001) Biology of freshwater pollution, 4th edn. Prentice Hall, London
MEA (2005) Ecosystems and human well-being: synthesis report. https://www.millenniumassessment.org/documents/document.356.aspx.pdf. Accessed 1 Nov 2017
Millennium Ecosystem Assessment (n.d.) Overview of the millennium ecosystem assessment. https://www.millenniumassessment.org. Accessed 1 Nov 2017
Monbiot G (2015) Do little, hide the evidence: the official neglect that caused these deadly floods. The Guardian, Tuesday 8 December 2015
Moors for the Future Partnership (n.d..) http://www.peakdistrict.gov.uk/looking-after/projects-and-partnerships/mff. Accessed Mar 2018)
Moss BR (2010) Ecology of fresh waters: a view for the twenty-first century, 4th edn. Wiley-Blackwell, London
Murray–Darling Basin Authority (2011) The living Murray story: one of Australia's largest river restoration projects. MDBA publication number: 157/11, Murray–Darling Basin Authority, Canberra
National Urban Forestry Unit (2010) Trees matter! bringing lasting benefits to people in towns. The Countryside Agency, the Rees Jeffreys Road Fund, and the UK Office of the Deputy Prime Minister, London
Natural England (2009) Green infrastructure guidance. NE176, Peterborough
Pitt Sir M (2007) Learning lessons from the 2007 floods. An independent review by Sir Michael Pitt. The Cabinet Office, London
Pötz H, Bleuzé P (2012) Urban green-blue grids for sustainable and dynamic cities. Coop for Life, Delft
Purseglove J (1988) Taming the flood. Oxford University Press, Oxford
Rotherham ID (ed) (2008) Flooding, water and the landscape. Wildtrack Publishing, Sheffield
Rotherham ID (2008a) Floods and water: a landscape-scale response. Pract Ecol Conser 7:128–137
Rotherham ID (2008b) Landscape, water and history. Prac Ecol Conser 7:138–152
Rotherham ID (2010) Yorkshire's forgotten fenlands. Pen & Sword Books Limited, Barnsley
Rotherham ID (2011) Chapter 9: the river Don as linear urban wildscape. In: Jorgensen A (ed) Urban wildscapes. Routledge, London, pp 131–140
Rotherham ID (2013) The lost fens: England's greatest ecological disaster. The History Press, Stroud
Rotherham ID (2014) Eco-history: a short history of conservation and biodiversity. The White Horse Press, Cambridge
Rotherham ID (2015a) Issues of water and flooding for trees, woods and forests. Arboric J – Int J Urban For 37(4).200–223
Rotherham ID (2015b) Bio-cultural heritage & biodiversity – emerging paradigms in conservation and planning. Biodivers Conserv 24:3405–3429
Rotherham ID (2017a) Recombinant ecology – a hybrid future? Springer Briefs. Springer, Dordrecht
Rotherham ID (2017b) Shadow woods: a search for lost landscapes. Wildtrack Publishing, Sheffield
Rotherham ID (2017c) The industrial transformation of South Yorkshire landscapes. In: Rotherham ID, Handley C (eds) The industrial legacy & landscapes of Sheffield and South Yorkshire. Wildtrack Publishing, Sheffield, pp 3–40
Rotherham ID, Harrison K (2009) South Yorkshire fens past, present and future: ecology and economics as drivers for re-wilding and restoration? In: Hall M (ed) Greening history: the presence of the past in environmental restoration. Routledge Publishing, London, pp 143–153
Rotherham ID, Harrison K (2012) Wilding by design as a future driver for a new nature in reconstructing South Yorkshire's Fens. In: Rotherham ID, Handley C (eds) Wild by design and ploughing on regardless. Landscape archaeology and ecology special series. Papers from the Landscape Conservation Forum, (3). Wildtrack Publishing, Sheffield, pp 54–72
Steffen W, Crutzen PJ, McNeill JR (2007) The anthropocene: are humans now overwhelming the great forces of nature. Ambio 36(8):614–621
The Economics of Ecosystems and Biodiversity (n.d.) Ecosystem services. http://www.teebweb.org/resources/ecosystem-services/. Accessed 2 Nov 2017

Thomas C (2013) The anthropocene could raise biological diversity. Nature 502:7
Thomas C et al (2004) Extinction risk from climate change. Nature 427:145–148
Ward-Thompson C, Aspinall P, Bell S (eds) (2010) Innovative approaches to researching landscape and health. Routledge, London
Watson R, Albon S (2011a) UK national ecosystem assessment: understanding nature's value to society. Technical Report. UNEP-WCMC, Cambridge
Watson R, Albon S (2011b) UK national ecosystem assessment: understanding nature's value to society. Synthesis of the key findings. UNEP-WCMC, Cambridge
Wilson EO (ed) (1988) Biodiversity. National Academy Press, Washington, DC

Abbreviations

AAD	Annual Average Damage
ABS	Australia Bureau of Statistics
ACT	Australian Capital Territory
AEP	Annual Exceedance Probability
AGWR-ADW	Australian Guidelines for Water Recycling–Augmented Drinking Water
AGWR-MHER	Australian Guidelines for Water Recycling–Managing Health and Environmental Risks
AGWR-SHR	Australian Guidelines for Water Recycling–Stormwater Harvesting and Reuse
AIDR	Australian Institute for Disaster Resilience
ANN	Artificial neural networks
ARY	Annual rainfall yield
ASEAN	Association of Southeast Asian Nations
ASR	Aquifer storage and recovery
ASTR	Aquifer storage, transfer and recovery
AWD	Annual water demand
BAPs	Biodiversity action plans
BASIX	Building Sustainability Index
BOD	Biochemical oxygen demand
BREEAM	Building Research Establishment Environmental Assessment Method
BS	British Standards
BTE	Bureau of Transport and Economics
CCC	Commission for Climate Change
CCW	Consumer Council for Water
CFMPs	Catchment flood management plans
CIRIA	Construction Industry Research and Information Association
Cl_2	Chlorine

V. Jegatheesan et al. (eds.), *Urban Stormwater and Flood Management*,
Applied Environmental Science and Engineering for a Sustainable Future,
https://doi.org/10.1007/978-3-030-11818-1

CORINE	Coordinated Information on the European Environment
CSIRO	Commonwealth Scientific and Industrial Research Organisation
CSO	Combined sewer overflow
CSOs	Combined sewer overflows
CSS	Combined sewer systems
CWs	Constructed wetlands
DAERA	Department of Agriculture, Environment and Rural Affairs
DARD	Department of Agriculture and Rural Development
DCLG	Department for Communities and Local Government
DECC	Department of Energy and Climate Change
DEE	Australian Department of the Environment and Energy
Defra	Department for Environment Food and Rural Affairs
DOC	Dissolved organic carbon
DWI	Drinking Water Inspectorate
DWTP	Drinking water treatment plant
EA	Environment Agency
ECA	Enhanced capital allowance
EDCs	Endocrine disrupting compounds
EU	European Union
FAO	Food and Agriculture Organisation
FME	Floodplain management entity
FPL	Flood planning level
FRM	Floodplain risk management
FRR	Flood Risk Regulations (2009)
FWMA	The Flood and Water Management Act (2010)
GI	Green infrastructure technology
GL	Giga Litre
GW	Grey water
HACCP	Hazard Analysis and Critical Control Point
HM	Her Majesty's
HR	Hydraulics research
ICA	Insurance Council of Australia
kL	kilo litre
kW	kilowatt
L	Litre
LDDs	Local Development Documents
LID	Low impact development
LLFA	Lead local flood authorities at local and regional scale
MAR	Managed aquifer recharge
ML	Million litres
NAO	National Audit Office
NFCERMS	National Flood and Coastal Erosion Risk Management Strategy for England
NFM	Natural Flood Management
NIEA	Northern Ireland Environment Agency

NPDES	National Pollutant Discharge Elimination System
NRA	National Rivers Authority
NSW	New South Wales
NT	Northern Territory
NTU	Nephelometric turbidity unit
NWQMS	National Water Quality Management Strategy
Ofwat	Water Services Regulation Authority
ONS	Office for National Statistics
PAHs	Polycyclic aromatic carbons
PCC	Per capita consumption
PECQ	Probable effect concentration quotient
PFAS	Per- and poly-fluoroalkyl substances
PMF	Peak maximum flood
PPCPs	Pharmaceuticals and personal care products
PPs	Permeable pavements
QLD	Queensland
RWH	Rainwater harvesting
SA	South Australia
SCC	Sheffield City Council
SEPA	Scottish Environment Protection Agency
SMP	Stormwater Management Plan
SPPS	Strategic Planning Policy Statement
SS	Suspended solids
SSS	Separate sewer systems
SuDS	Sustainable urban drainage systems
SWMP	Surface Water Management Plan
SRWH	Siphonic rainwater harvesting
TANs	Technical Advice Notes
TAS	Tasmania
TECQ	Threshold effect concentration quotients
UK	United Kingdom of Great Britain & Northern Ireland
UN DESA	United Nations Department of Economic and Social Affairs
USDE	Unit of Sustainable Development and Environment
USEPA	United States Environmental Protection Agency
UV	Ultraviolet
VIC	Victoria
VPS	Virginia Pipeline Scheme
WA	Western Australia
WFD	Water Framework Directive
WLC	Whole-life costing
WSPs	Water service providers
WSUD	Water-sensitive urban design
WWTP	Wastewater treatment plant

Glossary

Abstraction The removal of water from beneath the ground for the purposes such as irrigation

Annual rainfall yield The potential quantity of water that may be collected by a rainwater harvesting system per annum for a given type of design and location

Annual water demand The quantity of water used per annum

Anthropocene A term used to denote the current geological age, viewed as the period during which human activity has been the dominant influence on climate and the environment

Aquifer A subsurface layer of porous soil or rock that stores or allows for movement of groundwater or underground layer of water permeable rock such as chalk, providing storage capacity for water

Aquifer storage and recovery (ASR) An activity where water is transferred to an aquifer (by pump or gravity injection) for storage and subsequent extraction from the same point (e.g. for beneficial reuse)

Aquifer storage, transfer and recovery (ASTR) An activity where water is transferred to an aquifer (by pump or gravity injection) for storage and subsequent extraction from a different point (e.g. for beneficial reuse). Extraction from a point downstream may be undertaken to allow for treatment of stored water as it flows from the injection to the extraction point

Artificial neural networks Computer model designed to simulate biological neural networks

Attenuation tank Structure to retain water during periods of high water flow for later release in a controlled way

Biodiversity The richness and variety of life in the world or in a particular habitat or ecosystem

Biodiversity action plan Programme to identify and define species and habitats of conservation concern including the creation of targets for their conservation or restoration

V. Jegatheesan et al. (eds.), *Urban Stormwater and Flood Management*,
Applied Environmental Science and Engineering for a Sustainable Future,
https://doi.org/10.1007/978-3-030-11818-1

Bioretention A stormwater treatment practice which involves surface storage, vegetation, soil infiltration and water filtration

Bioretention systems/area Vegetated soil bed through which stormwater flows and is filtered during the process

Carbon sequestration The long-term storage of carbon in plants, soils, geological formations and the ocean

Catchment A defined area of land where surface water drains naturally and/or via human intervention to a single discharge point or the area over which surface flow contributes to a given location in a river or drainage system

Centralised supply system Water distribution via a large-scale supply network

Combined sewer overflows Combined sewer systems (see below for the definition) are designed to overflow occasionally and discharge excess wastewater directly to nearby water courses (streams, rivers, or other waterbodies). These overflows, called combined sewer overflows (CSOs)

Combined sewer systems Conveyance of both stormwater and wastewater through a single pipe network

Conceptual design First stage of a sustainable urban drainage system design process which maps local environmental features and the proposed development properties such as building type and land use

Constructed wetland A wetland designed and constructed for water quality improvement and water resource management

Conveyance swale Linear, wide and shallow depression covered with grass allowing the temporary storage of rainwater and the reduction of peak water flows

Critical infrastructure Assets that are essential for the functioning of a society and economy such as water supply, food production and its distribution, public health facilities, electricity generation, telecommunications, financial and security services

Cultural services The non-material benefits people obtain from ecosystems through spiritual enrichment, cognitive development, reflection, recreation and aesthetic experiences and thereby taking account of landscape values

Desalination Treatment processes which removes dissolved salts from a source water (e.g. brackish groundwater or seawater)

Detailed design Follows the conceptual and outline design stages which is a requirement of planning permission, tests and demonstrates the performance of the system

Detention pond Retains water temporarily for a short time period before entering a watercourse

Detention storage/basins Constructed storage positioned in a catchment to collect and temporarily hold runoff for peak flow mitigation. Discharge from a detention system proceeds downstream at a controlled flow rate

Diffuse pollution The release of pollutants over a broad area where the cumulative effect can be significant

Drainage infrastructure The network of underground pipes and chambers that in times of rainfall carry solely runoff, or a combination runoff and foul water, that is discharged to the nearest convenient waterbody (e.g. river), sometimes with little or no treatment. At higher return periods, roads may also convey water

Diverter rainwater harvesting system Similar to a total flow system, but only a proportion of the total runoff is diverted into a storage tank and the remainder bypasses to the drainage system

Dry (or enhanced) swales Linear wide and shallow depression covered with grass allowing the temporary storage of rainwater, situated on top of an underdrained filtration bed, designed to increase both the capacity of the system for flow attenuation and the level of pollutant filtration

Eco-cultural A description of landscapes and habitats which have developed through intimate and long-term interactions between people and nature

Ecosystem services The processes by which the environment produces resources utilised by humans such as clean air, water, food and materials. These services have been further categorised as provisioning, regulating, cultural and supporting services

El Niño A weather pattern of the El Niño–Southern Oscillation (ENSO) associated with a sustained period of warming in the central and eastern tropical pacific. This pattern often leads to dry/drought conditions over large parts of Australia

Evapotranspiration Surface water loss via direct evaporation in combination with transpiration from plants

Extensive green roof Green roof with a thin layer of substrate planted with hardy, slow-growing drought-resistant vegetation

Filter drain Linear trench filled with gravel used for intercepting and filtering flow from impermeable surfaces

Filter strip Gently sloping areas of dense vegetation (e.g. grass) designed to receive and remove sediment from runoff from adjacent impermeable areas or a strip of permanent vegetation designed to retard flow of runoff and thereby causing deposition of transported materials

Flash flooding Rapid flooding of a given area typically due to torrential rainfall

Flood mitigation Measures undertaken to manage floods to minimise their negative impacts

Floodplain management entity The government entity that has primary responsibility for managing flood risk

Fluvial flooding Flooding derived from a watercourse such as a river, caused by high rates of flow in response to e.g. heavy rainfall. Also, a flood event that is independent of an overflowing river or other water body, when heavy rainfall overwhelms an urban drainage system or when hillsides are unable to absorb water

Green infrastructure The network of vegetated (or green) places and associated hydrological systems to deliver multiple benefits including environmental, social, amenity and economic values to urban environments. Also an interconnected network of greenspace that conserves natural ecosystem values and functions and provides associated benefits

Green roof A vegetated landscape constructed on a roof surface designed for multiple reasons including, reduction of stormwater runoff volume, reduction in urban heat island, to provide spaces for people, as an architectural feature, to add value to property and to provide other environmental benefits. Also vegetation covered roof planted deliberately to reduce the flow rate of runoff and typically consists of layer of substrate planted with drought-resistant vegetation

Grey infrastructure A parallel concept to green infrastructure where the network of services such as roads, energy networks, water and sewage systems support the built environment

Grey water Wastewater from, e.g. sinks and baths in domestic or office buildings does not contain faecal matter

Groundwater Water present below the soil surface

Headroom methodologies The difference between the 'total water available for use' and 'water demand at any given time'

Hydromorphological Describes the physical characteristics of a water body, e.g. shape

Impervious surfaces Catchment surface area which rapidly produces runoff rather than allowing for infiltration. Typical examples include concrete, asphalt and block paving, and building roofs

Infiltration The process by which rainfall or runoff water moves through a pervious surface

Infiltration and detention basin Surface-based system with a depression in which water accumulates then infiltrates into the soil

Infiltration type rainwater harvesting system Collects rainwater in a storage tank from which an overflow infiltrates adjacent ground and acts as a 'soakaway', allowing water table recharge

Intensive green roof Green roof with a deeper substrate layer able to support a more structurally variable mix of vegetation; may be referred to as roof gardens

Intra-urban flooding The flooding of urban areas by runoff and/or mechanisms contained solely within the urban environment itself which includes surface runoff from roofs and pavements (pluvial flooding) or rising groundwater levels

La Niña One of the weather patterns of the El Niño–Southern Oscillation (ENSO) associated with a sustained period of cooling in the central and eastern tropical pacific. The La Niña event tends to increase rainfall over much of Australia

Low impact development (LID) A term typically used in the United States to refer to systems which are designed to reduce runoff and pollutant loadings by managing runoff close to its source(s)

Managed aquifer recharge The managed transfer of water including stormwater to an aquifer by pump or gravity injection for subsequent use and environmental benefit

Mire An area of wet, swampy or boggy ground

Outline design Uses information from the conceptual design stage adding additional detail on, e.g. expected runoff volumes; often a required component of a planning permission application

Parliamentary enclosures A series of UK Acts of Parliament that empowered enclosure of open fields and common land in England and Wales, creating legal property rights to land that was previously held in common

Peak flow mitigation The management and control of stormwater runoff to reduce risks from peak flows within drainage systems

Peat The brown deposit resembling soil, formed by the partial decomposition of vegetable matter in the wet and acidic conditions of bogs and fens

Per capita consumption Average consumption per person

Permeable pavement Paving surface that allows infiltration of water consists of impermeable parts, but with permeable joints; also paved surface areas which are designed to allow infiltration to occur into an engineered subgrade and/or native soil

Pervious pavement Pavement consisting of either permeable or porous paving

Pervious surface Catchment area which allows surface infiltration to occur. Typical examples include undeveloped open spaces (forests, grassland) as well as lawns, gardens and park spaces. Some pervious surfaces are deliberately designed to allow rapid infiltration of stormwater to subsurface soil

Pluvial flooding Flooding caused by surface runoff from roofs and pavements, typically in response to heavy rainfall

Porous paving Water infiltrates across entire surface, e.g. gravel, porous concrete

Potable water Water that is suitable for human consumption

Probable maximum flood The largest flood that could occur at a particular location. It will define the maximum extent of land liable to flooding and may be used in floodplain management plans

Provisioning services The products obtained from ecosystems, including food, fibre, fuel, genetic resources, biochemicals, natural medicines, pharmaceuticals, ornamental resources and fresh water

Rain garden Depression in the ground planted with vegetation has the capacity to collect and temporarily store stormwater runoff which then infiltrates into the surrounding soil over an extended period of time

Rainwater harvesting The collection of rainwater, predominately from roofs

Receiving (aquatic) environments Any environment downstream of a catchment (e.g. lakes, streams and coastal waters) that ultimately receives stormwater or other water runoff

Reclaimed wastewater Wastewater that is collected and treated to a fit-for-purpose reuse

Regulating services The benefits obtained from the regulation of ecosystem processes, including air quality regulation, climate regulation, water regulation, erosion regulation, water purification, disease regulation, pest regulation, pollination and natural hazard regulation

Retention and throttle-type rainwater harvesting system Uses additional storage (retention volume), which retains runoff during periods of high flow and then is emptied during low flow periods via a throttle valve

Retention storage/basins Constructed storage positioned in a catchment to collect and prevent discharge of a portion of runoff volume to downstream environments. Retention systems are designed for runoff volume reduction and (in some cases) peak flow mitigation and/or stormwater harvesting. Discharge from a retention system does not proceed downstream but is disposed of via infiltration or accessed for beneficial use

Retrofit Adapt or install new technology into an older structure

Reuse The collection and treatment of stormwater and wastewater for further beneficial use

Separate sewer systems Conveyance of stormwater and wastewater through separate pipework systems

Soakaway A chamber filled with material that will maintain its structure but contains a large proportion of empty space in which to store water; the stored water then infiltrates into the surrounding soil over an extended period of time

Sponge City A term used with reference to stormwater developments in China for enhanced flood control, water conservation, water quality improvement and natural ecosystem protection within the urban environment

Stormwater harvesting The collection of surface runoff from roads or pavements

Supporting services The services necessary for the production of all other ecosystem services, including soil formation, photosynthesis, primary production, nutrient cycling and water cycling

Surface drain Drainage system that carries surface water

Sustainable Use of resources (e.g. water) in a manner that will ensure continued availability into the future

Sustainable urban drainage systems (SuDS) A term used with reference to stormwater management in the United Kingdom incorporating natural approaches for stormwater drainage in urban environments and to minimise flooding and water pollution or systems designed to manage rainfall locally and mitigate flooding; used a variety of techniques and mechanisms including conveyance, run-off attenuation, infiltration and evapotranspiration

Total flow rainwater harvesting system Runoff enters the storage tank through a filter mechanism; excess water overflows from the storage tank into the local drainage system

Treatment trains The combination of stormwater treatment measures in series to improve water quality (e.g. to meet specified targets)

Twin-track approach Demand and supply-based water conservation measures and options used simultaneously

Upland Generally areas above 250–300 m in altitude

Urbanisation The gradual increase in the proportion of people living in urban areas and urban infrastructure development

Water-sensitive urban design (WSUD) A term used with reference to urban water management in Australia. The design is based on the total water cycle in urban areas where all available water resources are considered and integrated

Water service provider Entity responsible for the supply of mains water

Water stress level The demand for water relative to the supply capacity

Wet swales Depression specifically designed to contain wet/marshy conditions at the base; the mix of vegetation is purposely selected to increase the levels of filtration

Whole-life costing Costs for the entire anticipated lifespan of a development

Index

V. Jegatheesan et al. (eds.), *Urban Stormwater and Flood Management*, Applied Environmental Science and Engineering for a Sustainable Future,
https://doi.org/10.1007/978-3-030-11818-1

Printed by Printforce, the Netherlands